A Question of Biology

or

Life's Questions

"Ask one intriguing question, then answer it." This was the assignment given to four classes of Biology Honors students. Each student generated and refined one question. Each then sought the answer. This book is a collection of those questions and answers.

Genetics

What is the history behind the study of genetics?
Sarah Hartmus

Since the dawn of civilization, humans have possessed an inherent curiosity towards heredity. The earliest record of study in heredity begins with Hippocrates, a Greek philosopher who lived from approximately 460-375 B.C.E. He proposed that materials called pangenes were passed from a person to their eggs or sperm and then to a child, believing that traits developed during one's life, like scars could also be passed to children. Although this is incorrect, it was an impressive hypothesis for the time. Hundreds of years later, 19th century biologists studied ornamental plants' patterns of inheritance and correctly hypothesized that pangenes come from the mother and father. The pangenes then supposedly mixed and blended. This became known as the "blending theory" This was eventually proved false, as it leaves many gene and trait patterns unexplained, such as the way traits disappear and reappear in later generations.

"I am convinced that it will not be long before the whole world acknowledges the results of my work," prophesied Gregor Mendel in his journals. Later known as the father of genetics, he had truly hit the nail on the head. An Augustinian monk who lived in the mid-1800s, Mendel studied genetic patterns in the pea plants he bred. He was correct in believing that "heritable factors" (known today as genes) are passed from parents to children. Mendel discovered that genes remain independent. They do not mix and they can be moved and shuffled, but do not change individually.

With Mendel's groundbreaking work at his fingertips, Walter S. Sutton soon took the genetic stage with his study on the law of inheritance. Born in Utica, New York in 1877, Sutton studied Biology at the University of Kansas at Lawrence and later went into the field of medicine. Stumbling across Gregor Mendel's controversial work, Sutton was able to continue Mendel's studies and conclude that genes are stored in chromosomes. He published this idea along with his other research in his paper "The Chromosomes in Heredity" in the spring of 1903.

Working off of one another, Oswald Avery and Frederick Griffith were able to make the next breakthrough. Although they did not directly work together, their studies coincide well. They experimented with the pneumonia virus. The rough strain was put into lab mice. The mice showed no symptoms of illness. The smooth strain of the virus however did yield deaths. However, when both a denatured version of the previously deadly smooth strain and the innocuous rough strain were injected into the mice simultaneously, they became ill with pneumonia and died. This proved that something in the dead virus added to the harmless strain resulted in a live and very deadly strain. Avery and Griffith concluded that deoxyribonucleic acid (DNA) was the culprit and therefore was involved in heredity.

Through a series of experiments in 1952, Alfred Hershey and lab assistant Martha Chase were able to definitively prove DNA to be a genetic material. The two used the centrifugal force of a simple kitchen blender to separate viral from bacterial particles, proving phosphorus had entered cells and infected the nucleic acid. Their work helped to prove that DNA, not protein, is life's genetic material.

Erwin Chargaff's studies uncovered two rules which changed the course of genetic study forever. His work revealed the fact that the amount of thymine is equal to the amount of adenine and guanine levels are equal to cytosine levels in double stranded DNA. Chargaff discovered that these rules always prove true, although the exact amounts of the bases vary based on the organism.

The work of Rosalind Franklin in the lab of Maurice Wilkins was unjustly exploited by Watson and Crick. While working in the lab of Wilkins, Rosalind Franklin captured a remarkably clear x-ray of a sample of DNA which displayed its two strands. Wilkins, Watson, and Crick used this image without crediting her or her research. Watson and Crick won a Nobel Prize for their work in proving that DNA has two strands, neglecting to mention that it could never have been possible without Franklin's remarkable contributions.

For centuries, we have longed to know where we come from and why we are the way we are. Genetics play a major role in that equation. The rich history of the study of genetics has brought us the knowledge we possess today. Without the countless scientists who have devoted their lives to the pursuit of finding answers to questions of our genetics, we would be all the more in the dark about our very core and essence as humans.

***How do genes determine traits? (How does genotype affect phenotype?)**

Kendall McCaskill

Parents and grandparents pass genes from one generation to another. How are they expressed in people? Each person receives 23 pairs of chromosomes, in which lies the genotype. A genotype is the pair of alleles involved in a trait. This genotype affects phenotype, which is the physical expression of a trait. We will see how genes are expressed on a molecular level as well as things that affect it.

An organism's genotype contains sequences of nucleotide bases that are in DNA, that carry sequences that code for one's traits. DNA is translated and transcribed, which is the first step of gene expression. DNA, deoxyribonucleic acid, is genetic material found in all living organisms that carries information ultimately used in the development of organisms. First, during transcription, DNA is copied and rewritten in terms of RNA (messenger RNA) by an RNA polymerase (in respect to base pairing). The second part of gene expression, translation, occurs in three steps, initiation, elongation, and termination. After mRNA is synthesized, the first step, initiation begins. This is when the message previously transcribed is sent to the ribosomes, an organelle in which DNA is translated to proteins. mRNA is read in triplets called codons. This genetic code determines how they are translated into an amino acid sequence. The second step, elongation, consists of amino acids being brought to the ribosomes and are linked together to create polypeptides. The last stage, termination, occurs when the ribosome encounters a stop codon, a signal to stop, releasing the polypeptides.

The sequences of DNA are transcribed and translated in order to create proteins that further determine phenotype. Several genes code for proteins that are involved in the expression of the phenotype. The base pairings determine how amino acids arrange to further form a protein. Often, several genes provide information for synthesizing different genes that affect a single physical trait. Very common genes and proteins include MC1R, the gene that controls the production of the pigment melanin. This pigment is found in skin, hair and eyes, and these are also all affected by an individual's genotype. Other genes like GHR provide instructions in creating growth hormone receptors. Physical traits are widely dependent upon the synthesis of a protein, yet others are caused by mutations and are affected by the presence of one gene.

Some physical traits are affected by the presence of a single gene. Traits like blood type are Mendelian traits, traits that are affected by the presence of a single gene. Traits are passed down in the form of dominant and recessive alleles. These two common terms are both used in the case of inheritance, alleles being a form of the gene, and one being shown when being dominant over a recessive allele (and is usually not present). For example, the protein RHD decides the antigens in blood produced, which the determines one's blood type being A, B, or O. In some cases, the presence or absence of a gene can negatively affect an organism. Any alteration in a single nucleotide sequence is called a mutation. Mutations in the phenotype of an organism can alter health, appearance, and make them more susceptible to different diseases. Mutations can occur in substitutions, eradication, as well as insertion. These can all change, in turn creating different amino acids. Common disorders include sickle cell anemia, a disease linked to the alteration of one amino acid. Other common genotypes include albinism, dwarfism (achondroplasia), cystic fibrosis, and Huntington's disease.

Other physical traits are affected by the presence of several genes. Many phenotypes present in humans are not based on a simple gene but are affected by the presence of several genes.

As previously mentioned, proteins are synthesized through the process of transcription and translation that affect the phenotype. Traits like eye color, hair color, and skin color are determined by the presence of more than one protein. This is called polygenic. Eye colors' most important proteins include OCA2, HERC2, TYR, SLC24A4, and SLC45A2. Skin color is also controlled by several genes and like others, is based on ancestry.

In conclusion, proteins play a crucial role in determining the phenotype of an organism. Proteins are synthesized through the process of gene expression, including transcription and translation. Some physical traits are affected by only one gene and in some cases, possibly a mutation. Polygenic traits are based on the presence of more than one gene.

***What are the genetics of cancer?**
Katie Hopper

Cancer is a disease that can affect most parts of the human body. But what exactly do genes have to do with cancer? Genes play an important role in developing cancer and in the susceptibility of an individual to certain kinds of cancers. Genes that mutate to form cancer are called oncogenes and Tumor-Suppressor genes. An individual's genes can also affect what types of cancers can develop. These genes are inherited from parents.

Oncogenes are one of the two types of genes that are mutated to become cancerous. Before mutation, Oncogenes begin as proto-oncogenes which are already located in the body and prone to becoming cancerous. Oncogenes can be activated by a change in an amino acid which can alter the shape of the protein and cause the proto-oncogene to become mutated. Some oncogenes are also activated when a gene is rearranged between chromosomes. An example of a cancer caused by this rearrangement is Chronic Myeloid Leukemia. The Chronic Myeloid Leukemia oncogene is activated with the chromosome rearrangements of the ABL gene. Both methods of mutation cause a change in multiplying patterns, causing the mutated gene to multiply and spread erratically. In summary, oncogenes are a type of gene mutated to become cancer.

The other type of genes that are mutated to become cancerous are Tumor-Suppressor Genes. Tumor-Suppressor genes require a loss of a gene from a cell to become mutated. The gene can also break off for it to become cancerous. An example of a Tumor-Suppressor gene is the BRCA1 gene, which everyone is born with. This gene regulates the rate of cell division while also repairing damaged genes. BRCA1 can malfunction and become cancerous, specifically as a variant of breast cancer. Tumor Suppressor genes are found in both hereditary and sporadic cases of cancer, but they are also harder to identify during early stages of mutation.

Hereditary cancers are cancer genes passed down through a person's family. Every type of cancer can be inherited but some are more commonly inherited than others. An example of a commonly inherited cancer would be colon cancer. Around 5-10% of common cancers are caused by genetic susceptibility due to a single dominant inherited gene. Susceptibility to rare familial cancers is also a dominant trait. The chances that the gene inherited become cancerous are low and usually depend on age. People can help prevent the risk of cancer by being screened for inherited genes. A family history of certain types of cancer also help with the testing process for these inherited genes. The downside is that scientist still have trouble identifying those at risk for rarer familial cancers.

In conclusion, genes play a vital role in a person's susceptibility to cancer and how cancer develops. The genes that help with cancer development are oncogenes and tumor suppressor genes. These genes are important in the process of creating cancer as they mutate a healthy gene. Both these genes are found in hereditary cases of cancer. These inherited genes are not cancer, but are more prone to developing cancer. Understanding the genetics of cancer is vital as it can help with prevention and maybe even developing a cure.

***What effects do GMOs have on humans?**
Ben Walter

GMOs have been a topic of hot debate over the last couple of years, even leading to the creation of a whole new market for non-GMO food. But how do these seemingly new inventions affect humans? In order to answer that question, four things need to be done: Define what a GMO is, see the evidence against GMOs, look at the evidence for GMOs, and weigh them.

GMO simply stands for genetically modified organism. What most people don't realize is that the term also applies to ancient human agriculture, when people selected seeds from the biggest corn to plant, also known as selective breeding. However, in order to keep things simple, the term GMO will be used to describe organisms who, through a laboratory process, had genes of other organisms forced into their own. In plants, the goal of this process is usually to give the plants a favorable gene, like insect resistance, that they don't normally have.
The pros of GMOs are obvious. According to UN data, one child dies every 7 seconds due to starvation or hunger attributed causes. GMOs increase the efficiency of growing crops, meaning that fewer precious resources, like water, have to be used to attain larger yields. GMOs also help areas of the world with vitamin deficiencies. One example of this is Golden Rice, a crop engineered to have high amounts of vitamin A to help the 500,000 children who go blind every year. Not only does the introduction of GMO plants help solve world hunger, but they also are good for processed products. Since the processed products of GMO's, like corn oil, are chemically the exact same as non-GMO corn oil. This shows that even if used for just processed products, GMOs can be far more efficient without losing any quality of final product.

Although the good GMOs do is obvious, the cons are a little less clear. Although there are fears that GMOs cause cancer and other deadly, harmful diseases, there is no firm scientific proof of it. Possibly the most popular study claiming that GMOs cause cancer is by Gilles-Éric Séralini. He claims to have fed GMOs to rats, and the rats, in turn, got cancer. However, this study has been debunked and rejected by the scientific community because of it's cherry-picked data, a small sample group, and the fact that Séralini used a strain of rat that had a 80% chance of developing cancer in its lifetime. Although there are no proven physical effects of GMO's on humans, there are both economic effects and religious conflicts. Only major companies that have large budgets have the resources to develop GMOs. This can, and has already, led to monopolies in this new field. Others argue that it is unethical to the consumer to sell them GMO foods.

The debate over GMOs comes down to the number of people it helps. The introduction of GMOs has the potential, and has already saved millions of people from hunger. When compared to the relatively flimsy evidence against it, the effects of GMOs on the human race in definitively good. With the world population growing rapidly, and the amount of resources shrinking quickly, we need a solution to feed future generations. Although we will need to make sure to closely monitor these companies and development, the future seems to be full of genetically modified food.

***What is the role of recessive genes and alleles in genetic disorders?**

Ji Hoon Choi

Some symptoms of diseases can be found just by looking at a person, which is also true for many genetic disorders. But what if someone is a carrier of a disorder and they do not know? How would they know if their offspring is affected by the disorder? The Mendelian explanation and the use of a punnett square helps shed some light to the questions of genetic recessive disorders.

One way to explain how these types of genetic disorders get passed down is from the Mendelian explanation, the backbone example for recessive traits. Discovered by Gregor Mendel, the explanation states that certain traits are dominant, meaning they are more likely to be displayed than the recessive genes. These two forms are termed alleles. Alleles are the different versions of a gene, such as the allele for blue eyes for the eye color gene. The dominant traits show physically as a phenotype in a heterozygous person (one possessing a recessive and dominant allele). For example, the eye color brown is dominant over blue, so it is likely that if a parent with brown eyes and a parent with blue eyes had an offspring, the offspring will have brown eyes as their phenotype. So, how do some people show up with recessive disorders if the disorder is recessive? The answer to this is like how a child can have blue eyes. The offspring will need to be homozygous (two of the same allele) of the trait/disease for it to show up physically.

There is a way to predict the percentage of likelihood for a certain trait. This method is known as the Punnett square. A Punnett square is designed as a square divided into four parts, each part has a gene from a parent. The top of the square will include alleles of one of the parent while the side depicts the other. It is set up like the lattice method in multiplication, except the numbers in the lattice method will instead be the alleles that the parents carry. Inside the smaller squares will be the result of the allele combination of an offspring. If done in the correct way, for heterozygous parents carrying a recessive disease allele each, the result will be a 25% likelihood of a child with the disorder showing through as the child will be homozygous.

Another way recessive disorders can be passed is by x-linked chromosomes. An example of this would be the Romanovs of Russia. The mother was the carrier of a disease known as hemophilia on her x-chromosome,but wasn't affected as the other x-chromosome needed for all females blocked the disorder from surfacing. The Romanovs had four daughters and a son. For the daughters, they also became carriers as one x-chromosome from their mother was affected, but the the affected chromosome was blocked by the healthy x-chromosome given from their father. But it was different for the son, as he had the affected x-chromosome and a y-chromosome from his father, leaving him no way to block the disease. This caused him to have the effects of hemophilia.

Even with genetic recessive disorders, there are still possibilities to avoid and know the disorders from surfacing physically. We saw this through how recessive traits in general are passed down to offspring, seeing the Punnett square, the percentages of carrying the disease is definitely higher than being directly affected by it. Thus, thankfully, recessive genetic disorders, like the name states, does not surface very often.

***What is the difference between dominant and recessive genetic disorders?**

Nikita Shishu

A genetic disorder, also called an autosomal disorder (because twenty two pairs of the twenty three pairs of chromosomes are called autosomes), is a disorder that is passed on from one generation to the next generation through genes. An allele is an alternative version of a gene. Organisms inherit two alleles for each trait; one allele is from their mom and one allele is from their dad. An organism that has two identical alleles for a gene is homozygous and an organism that has two different alleles for a gene is heterozygous. Often, if the two alleles of an inherited pair are different, one determines the organism's appearance and the other one has no noticeable effect on the appearance. Sperm and eggs carry one allele for each inherited characteristic. Genetic disorders can either be dominant or recessive.

Autosomal dominant genetic disorders are caused by dominant alleles. A person only needs to inherit one dominant allele to get a dominant genetic disorder. In this case the healthy gene is recessive and the gene with the genetic disorder is dominant. If the person inherits two dominant alleles, they will have the severe version of the genetic disorder, and will most likely die early. They will most likely not get the chance to become an adult. If the person gets one dominant allele and one recessive allele, they will still have the genetic disorder. However, the disorder will be moderate, and the person will not die as early. If one inherits two recessive alleles, they will be perfectly healthy. An example of a well-known dominant disorder is Huntington's disease, which can cause muscular degeneration and dementia.

To inherit a recessive genetic disorder, a person must inherit two recessive alleles. The gene with the genetic disorder is recessive and the healthy gene is dominant. If the person gets two dominant alleles, they will be completely healthy. If the person gets one recessive allele and one dominant allele, they will carry the genetic disorder in their genes, but there will be no noticeable effect on the person. If the person gets two recessive alleles, they will be a sufferer of the genetic disorder. An example of a recessive disorder is cystic fibrosis, which causes lung problems like pneumonia. The chances are lower of getting a recessive disorder. Also, most genetic disorders are recessive.

In conclusion, genetic disorders are passed on through a person's DNA. The sperm and the egg carry the genes, or alleles. The genetic disorder can be dominant or recessive. Only one allele is required to get a dominant genetic disorder, and two alleles are required to get a recessive genetic disorder.

***How have DNA profiling techniques developed over time?**
Jason Cho

Since its introduction, DNA profiling has had a massive effect on the world of forensic science, but where did the process originate from and how has it evolved? Additionally, where did some of the main processes used in DNA profiling come from, and how were they discovered? How has DNA profiling evolved since its introduction? This essay will aim to answer these questions and more in order to shed light on one of the most important aspects in forensic science and in the world of law.

One of the main processes in DNA profiling, Gel Electrophoresis, was developed long before DNA profiling. Gel Electrophoresis is a branch of electrophoresis, the process of separating biomolecules using an electric current. Electrophoresis was introduced in 1937 by Arne Tiselius, a Swedish chemist, and was original used to separate proteins, but has evolved to separate DNA, RNA, and other biomolecules. Later, it was discovered that putting the biomolecules in different materials and using an electric current would separate the biomolecules as well, leading to use of gels in electrophoresis.

The introduction of the Polymerase Chain Reaction helped soon introduce the world to DNA profiling. Polymerase Chain

Reaction, PCR for short, is the technique that amplifies the number of specific DNA sequences and is essential to any DNA analysis. The discovery of PCR is credited to Kary B. Mullis, who conceptualized it in 1983. Over the next seven years, Kary would continue to improve PCR, which eventually became more efficient than having bacteria replicate DNA. Eventually, this process would be used to help change the world of law and forensic science.

DNA Profiling was finally introduced to the world in 1986, by a man named Alec Jeffreys. In 1984, Alec Jeffreys, a geneticist working at the University of Leicester, discovered a patch of DNA that was unique to everyone. It wasn't until 1986 that Alec would use this knowledge to convict Colin Pitchfork, with murder and rape. With Colin's conviction, DNA profiling became an integral part of the world of forensic science and justice.

Despite being introduced only 32 years ago, DNA profiling has undergone some changes. The first instance of DNA profiling was not perfect and it was clear modifications could be made. After the conviction of Colin Pitchfork, many companies became interested in the process, and would soon start using DNA profiling in their business and cooperate with each other in order to improve the technology. In the modern era, the methods originally used have changed significantly. From using the Southern Blot technique to Polymerase Chain Reaction, replacing slab gels with capillary electrophoresis, and making the process more automated.

Arne Tiselius developed electrophoresis, which would eventually offshoot gel electrophoresis. Kary B. Mullis introduced PCR to the world, and Alec Jeffreys would revolutionize the forensic science scene with his introduction of DNA profiling. Thanks to the efforts of these three people, DNA profiling has made its way into forensic science and has become essential in correctly convicting suspects.

How do we analyze DNA and how does it contribute to forensic science?
Amelia Gold

The study of scientific evidence for crime scene investigations is known as forensics. DNA profiling is crucial for forensics because of its ability to place DNA samples with certain people, such as suspects in a crime. In order to test these DNA samples, scientists look at the sequences in the genomes, called genetic markers, which are different in every person. In order to investigate a DNA profile from a crime scene, DNA must be taken from the crime scene, amplified to create the DNA on a larger scale and then compared.

One way we are able to amplify samples of DNA is with PCR, or polymerase chain reaction. PCR is used to replicate a DNA sequence from very small tissue samples in a test tube. It amplifies DNA in order for scientists to analyze it closely. In order for copies to be made of the DNA sequences, the reaction must be cycled through a series of temperature changes repeatedly. By repeatedly denaturing and synthesizing a single DNA strand using heat and the enzyme Taq polymerase, copies of the original DNA strand can be made. This method easily replicates STRs, or short tandem repeats. These are repeating DNA sequences that are common DNA markers, varying in size. DNA that is amplified by PCR can be then looked at for gel electrophoresis, sequencing, or cloned for further experiments. This technique can be used in many areas of biology along with scientific forensics.

Gel electrophoresis enables us to break down and analyze DNA samples. Electrophoresis is how particles in a fluid/gel move while being influenced by an electric field/current. It separates the molecules of fragments of DNA in order to be further analyzed. Only minimal amounts of the initial DNA samples are needed, and they can even be in a somewhat damaged state. The DNA is placed into a specific gel called agarose, which is a polysaccharide. DNA is placed in "wells", or holes on one side of the gel. When the electrical current is sent through the gel, because DNA is negatively charged, the DNA will be attracted to the positive side and will separate out into varying strands. The longer strands move slower and will not travel through the gel as quickly as the shorter strands, leaving them all individually separated.

STR analysis is the comparison of the lengths of short tandem repeat sequences (STRs) in certain locations in the genome. Scientists use data about specific sets of genetic markers to create DNA profiling. Repetitive DNA are multiple copies of DNA sequences found in the genomes. These very short DNA sequences repeated many times are called short tandem repeats. The type of sequence, where it is located, and how many times it repeatsare different with every person.

DNA profiling can produce evidence in forensic investigations. If a crime is committed, pieces of tissue or body fluids can be left at the scene of the crime or found on the clothes of the victim or the aggressor. For example, in the instance of rape, semen from the assailant can be recovered from the victim. Because everyone has a unique DNA sequence, DNA profiling can reliably and accurately match DNA samples with the person. Also, methods such as gel electrophoresis and PCR can be used to replicate/amplify very small samples and even damaged samples of DNA.

What are some uses of genetic markers?

Mei Ling Wood

Genetic markers are short gene or DNA sequences that can be used for many purposes. Scientists can sample DNA with single copies of their genetic markers, which can be used to conserve endangered species. It is also commonly used in forensics for identifying of genes involved in inherited disease and differentiating DNA sequences to identify individuals and populations.

As stated above, genetic markers are short gene or DNA sequences with a known location on a chromosome, where the markers vary from person to person. It can be used to identify specific individuals or species and all the genes inherited on a chromosome by comparing the markers of the individuals. Single nucleotide polymorphisms (SNPs) and minisatellites are examples of genetic markers. SNPs represent all the variations of gene or DNA in a nucleotide, while minisatellites are tracts of DNA that are repeated from 5-50 times.

Restriction fragments length polymorphism (RFLP) is a difference in DNA sequences as restriction enzymes cut specific parts of the DNA. Restriction enzymes cut DNA at restriction sites (location on DNA that are recognizable to restriction enzyme), causing the DNA sample to be broken up into fragments. The resulting fragments will then be separated according to their length by Gel Electrophoresis, revealing unique patterns to a specific part of the chromosome, which will differ for those tested. RFLP analysis can be used for DNA fingerprinting, identifying suspects based on samples of evidence that are collected at the crime scene. It can also be used for tracing ancestry and paternity, studying evolution of wildlife, and detecting certain diseases.

Tandem repeats are repeating DNA patterns with one or more nucleotides adjacent to each other. Even though the full genetic profiles of humans are the same—repeated sequences of DNA are identical—the number of the repeats differ. Stretches of DNA repeats are known as variable number tandem repeats (VNTRs). These short nucleotide sequences are often located in the intergenic regions, and are basically long STRs, each barring 10-100 base pairs. When evidence samples from crime scenes are sent to labs, VNTRs are removed from the surrounding DNA by the Polymerase Chain Reaction or RFLP methods, which then undergo Gel Electrophoresis to get ready for testing. Similar to RFLP analysis, VNTRs can be used for the study of genetic diversity, and distinguishing bacterial pathogen.

Short Tandem Repeats (STRs) are considerably the most widely-used genetic marker in the forensic field as it’s length of repeating DNA sequence at a given loci (position on a chromosome) varies from person to person. Similar to VNTRs, STRs are regions of repeated nucleotides sequences of DNA, but the length is limited to two to ten base pairings; they also appear more frequently in the genomes than VNTRs. For example, two people can share DNA GATAT, but one person can have that DNA repeated for 13 times, while the other only have 8 repeats of it at a given loci. The FBI uses up to 13 different STRs for individual DNA profiling as the number of matchings between the given marker and suspects reduce when multiple STRs are taken- the probability can reduce from 1/10 until 1/10 trillion, which can be considered as a definite match.

Genetic markers have become essential in science, and as time goes on, new genetic markers are being made. While RFLP, VNTRs, and STRs are several genetic markers commonly used today, the thousands of other markers are also used in several different fields.

How do enzymes assist in DNA replication and transcription? What specific components make this process what it is?

Jovun Dhillon

There are multiple steps when replicating DNA, which involves many components like the synthesis of strands and coils by replication enzymes. Replication is the process in which two strands of DNA are copied to produce two identical strands. For these processes to occur, the existing strands of DNA act as templates of replication. Steps one and two are the separation of two strands of DNA and the replication fork formation which results from helicases. Then, long chains of nucleotides are built using base pairing.

We will go in depth about each specific step in the first part of replication, the separation of two strands that make up a molecule of DNA. The separation of these two strands will give way to the helix that needs to be copied. Helicases are a class of enzymes that are involved with the separation of DNA strands that make up the original helix. Helicases unwind coils of DNA by breaking apart light hydrogen bonds that connect DNA down its middle. Helicases are energized by hydrolysis of ATP. The helicases simply “melt” the bonds to create the open strand that is to be copied. Once the strand separation is completed, DNA polymerase comes in. DNA polymerase is an enzyme that works at making new strands of copied DNA, by synthesizing new chains of nucleic acids using the base pairing rules. This process eventually leads to the formation of two identical strands from one original DNA coil. Helicases tend to start the breaking process at replication points, where a for formation will occur from the separation of two sides parental DNA.

Once the DNA strands are separated, enzymes participate in preventing supercoils. Topoisomerases are a class of enzymes that participate in the unwinding of a DNA strand. This process prevents supercoils, which are DNA helices that are twisted tighter and tighter. If you have a string and twist it, it will eventually be wound so tight, that it will curl and wrap around itself. This is how topoisomerases function in the replication process. Once the strands of DNA are separated and new base pairs are produced, DNA ligase comes in. Ligase is an enzyme that joins the segments of copied strands of DNA by creating a new bond. One responsibility of ligase is to connect copied strands and to repair the DNA by joining Okazaki fragments (newly formed DNA fragments) formed during replication. The strands are connected by a covalent bond in the sugar-phosphate backbone. As a result, a chemical bond forms via hydrolysis from a chemical group.

DNA replication involves multiple classes of enzymes. This type of replication gives way to how DNA is passed and copied through generations. To summarize, the helicases break apart DNA while polymerases create chains of nucleotides. After the fork formation, topoisomerase prevent supercoils and ligases join new sections of DNA. These processes are vital to the survival and growth of all organisms on earth.

***What are the benefits of Genetically Modified Organisms?*

Eric Gordy

The use of DNA recombinant technology to create Genetically Modified Organisms (GMOs) creates vast benefits for the global population, such as vaccines against diseases, insect resistance for plants, and the fortifying of crops with nutrients. For all GMOs, scientists combine pieces of DNA into different sources. This DNA can be used in animals, plants, bacteria, and other organisms, and combining certain pieces of DNA creates extremely beneficial results that can aid in sustaining the human population.

The use of GMOs in the field of medicine can create both important vaccines and hormones. In terms of vaccines, scientists produce large amounts of the protein found on a certain pathogen's surface, which is the main ingredient for the vaccine of that pathogen. DNA recombinant technology creates an innocuous artificial mutant of that pathogen, which, when inserted into one's body, creates a strong immunity to this disease organism. This technique has been notably used to create the modern smallpox vaccine. Along with creating vaccines, GMOs can be used to develop bacteria that secrete insulin. These bacteria, when inserted in one's body, provide the body with a constant source of insulin, which is critical for insulin-deprived people. GMOs can also be used to create HGH, or the Human Growth Hormone, by combining a human DNA fragment to a chemically created piece of DNA with *E. coli*, creating HGH high in *E. coli*. This allows the HGH to be inserted into humans with growing disorders, such as dwarfs, giving them a better chance for growth. GMOs not only have an important role in the realm of medicine, but also in farming and plants.

In many plants, the use of GMOs can help to create insect-resistance and disease-resistance. Bt, a gene from soil bacteria, can be added to plants using DNA recombinant technology. This gene induces the plants to secrete a protein that is toxic to insects, decreasing the need for chemical insecticides (eliminating their cost), increasing crop yield, and saving farmers an average of $2.80-$14.50 per acre. GMOs can also be used to create versions of crops resistant to many diseases, increasing crop yield by decreasing the number of diseased plants. GMOs use in plants not only increases plants sustainability, but also can reinforce plants with nutrients.

The up-and-coming use of GMOs is the fortifying of plants with nutrients such as vitamins or proteins. DNA recombinant technology can serve a critical role by supplying traditionally nutrient deprived peoples with proper nutrients, and in some cases removing unhealthy proteins, starches, oils, or other biomolecules from foods. Genetic engineering can be used to transfer nutrients to crops, strengthening them with beneficial nutrients, and providing plants with higher levels of energy. One important example of this is the production of golden rice, which is injected with carotene-rich Vitamin A. This has been produced in many Southeast Asian countries, such as Bangladesh and the Philippines, nourishing historically nutrient-deprived groups. The rice is usually available for little or no cost, showing its design to remove the financial burden and help the consumer. Golden rice production is an example of a technology transfer program, which provides genetic engineering technology to poor countries at little to no cost, giving countries easier access to nutrient-fortified crops. Many other products like golden rice exist in the world today, showing genetic engineering's important role in sustaining the human population.

As evidenced by the vast impact of the many uses of GMOs, DNA recombinant technology can be a powerful and positive force for the human population. Genetic engineering can be used to synthesize important vaccines and hormones to improve human health. It can be used in plants to create insect and disease resistant plants, increasing crop yield while decreasing cost for the farmer. GMOs use in plants extends to the creation of nutrient fortified crops, fueled by technology transfer programs, which provide GMOs to developing countries. Genetically Modified Organisms, as shown by their use in human medicine and plants, are truly important benefactors in sustaining a healthy world.

***How do Genetically Modified Organisms affect agriculture?**
Nina Govila

Genetically modified organisms can be used in the production of foods which can be harmful to humans if consumed. This concept has been continuously brought up in the debate if genetically modified organisms should be used in our modern agriculture. There are many fears and false claims that genetically modified organisms can negatively affect our health, although these fears have never been proven. There is no evidence to the statements that genetically modified organisms can harm humans,

Genetically modified technology can help many people and prevent viruses and diseases from getting into our crops. The cassava crop, banana crops, golden rice, corn crops, and many more were all affected with viruses, and they could all have been fixed by genetically modified organism technology. However, it is being withheld from farmers because of many voices of paranoia and fear.

The anti-science movement is the biggest threat to global food security. In 2013 alone, 3,100,000 children died due to malnutrition, while no children were killed because of genetically modified organisms. With the use of GMOs in agriculture, crops could be produced at a much faster rate, and would be able to reach more households with a lower income. This could save children from having to suffer from malnutrition, as well as being less expensive to the overall economy.

****What is the rationale behind bioluminescence gene transfers to other organisms?***

Alexandra Miller-Thomas

Scientists have created bioluminescent, or glowing, animals in order to find a functioning gene to immunize animals against specific diseases that humans are also susceptible to. By combining the immunocompetent gene (uncompromised) with a bioluminescent (glowing) gene, scientists are able to test the function with visible results. Without the combination of the two, the only way to test the functionality would be by injecting the specific pathogen into the animal which could lead to the death of an animal whose immunocompetence was impotent. Using the process of injecting the gene duo into trial embryos, scientists have been able to test different types of immunocompetent genes on multiple animal species. These tests, revolutionary to the testing, are curing many diseases of humans and other animals.

In order to test the functionality of a immunocompetent gene without physically inflicting the disease on animals, scientists interlaced the immunocompetent gene to a bioluminescent gene that gives off a visible sign of functionality. This not only helps the scientists with finding the functioning genes out of all trials, but the bioluminescence also helps save the lives of the animals who were the trial subjects. The safer the trials are, the more likely the scientists are to receive funding for more test trials. This is important because these animals are being tested with diseases that are not usually transmitted to humans. By keeping the animals safe and increasing the number of tests able to take place, scientists have created a revolutionary way of testing which can one day become a way to cure many diseases.

Using the visible results of the bioluminescent gene, scientists found that the animals that glowed under a blacklight were also immune to the specific disease they were immunized for. The interlaced genes are combined to a point where if one shows functionality, the other will function too. Had the connection been weak, the scientists could possibly end up with bioluminescent animals that are not immunocompetent or vice versa. This is a problem because the immunodeficiency is the goal but if the connection is weak and there is no way to tell which gene functioned properly, there is no point in connecting them in the first place.

The gene duo was injected into the embryos of the trial animals' mothers. This process inserts the gene into the unborn eggs of the test subjects allowing for the gene to become part of their chemical makeup. Depending on whether the gene was rejected or encompassed by the embryo, the trial subject will either become immunocompetent and bioluminescent, or neither. They inject the gene duo because the results are most promising when the gene has a longer period of time to connect the the chemical makeup of the embryo rather than a one day old animal whose chemical makeup is finished for the most part.

Though many trials have worked on animals such as cats, a trial done on monkeys at the Oregon Regional Primate Center was not as successful. This trial involved 20 gene-injected embryos and four births. One of the births contained two immunocompetent and bioluminescent twin monkeys who miscarried. Out of the other three monkeys born, one was immunocompetent and the other two were neither immunocompetent nor bioluminescent. This trial was one of many failing trials due to the results not containing any living immunocompetent and bioluminescent animals.

Through many trial and error tests, scientists were able to find a stable immunocompetent gene by intertwining it with a bioluminescent gene in order to help test the functionality.

How do genes interact to form a phenotype?

Will Withrow

Many people wonder how and why they look and function the way they do. It comes from the genes which interact to form what is known as a phenotype. Many different factors play a role in determining an individual's characteristics, and they are all controlled by the genes coming from parents.

A person's genes are inherited. These genes determine an individual's traits. Chromosome pairs are formed by a combination of one chromosome from the mother and one from the father. The interaction, dominance or recessivity determine traits such as eye color.

Phenotype is the scientific word for the expression (appearance) of a trait. A phenotype can be anything from hair color to leg length. It is determined by genes working together. A "gene" is a component of a chromosome and comes from one's parents. Different genes on different chromosomes determine various phenotypes. Each chromosome pair represent one from the mother one from the father and they determine an individual's traits. Different genes on different chromosomes determine various traits. The twenty-third chromosome pair, for instance, determines a human's sex.

Although all of an individual's genes come from parents, some are more likely to be expressed than others. This is the dominant form (or allele). If each parent has one dominant and one recessive gene, there exists a 3 out of 4 chance of having the dominant phenotype because in order to express the recessive trait one must receive a recessive gene from each parent. Often, people with dominant traits actually have both a dominant and recessive gene but do not express the recessive phenotype. But there is a chance that their children could receive the recessive trait. We see this in dominant genes such as brown eyes.

The study of genes had led us to gain a greater understanding of how traits are passed down through the generations, and how to predict the traits of individual's offspring. Almost every aspect of an individual's appearance comes from the parents and each aspect is its own phenotype which explains why children often look like their parents. A person's traits aren't so random after all.

How can gene therapy be used to treat disease?

Avaneesh Prasad

Gene therapy is a revolutionary aspect to modern medicine. Scientists have been discovering new ways to treat diseases with this every day. There are several types of gene therapy, each to treat different types of diseases. There are also a couple different approaches to gene therapy, each effective for different problems. With more and more genes being mapped, gene therapy can become more efficient and effective.

Gene therapy can be effective for monogenic diseases, polygenic diseases, and infectious diseases. In a monogenic disease, one gene is not working, which results in the disease. In a polygenic disease, multiple genes are defective, making it harder for gene therapy be used to effectively treat the diseases. Infectious diseases are the hardest to cure with gene therapy because it is the result of infectious agents multiplying inside the body.

Researchers are testing three main approaches to gene therapy, which include replacing a mutated gene with a healthy copy of the gene, deactivating a mutated gene that is incorrectly functioning, and inserting a new gene to help combat a disease. Introducing a new gene into the body is the main approach to gene therapy. Researchers have genetically engineered viruses called vectors to carry the new gene. There are two types of vectors: retroviruses and adenoviruses. In germline gene therapy, retroviruses put the new gene into a chromosome in the human cell. In somatic gene therapy, adenoviruses put the new gene elsewhere in the nucleus and are used for somatic gene therapy.

Somatic gene therapy is one version of gene therapy, where a modified gene is inserted into the patient to treat the patient's condition. The transfer of genes happens in places like bone marrow, where new DNA will not be affected by it, so the changes will not enter the sperm or the egg. Copies of a normal human gene are placed in a bacterium. The bacteria can be stored in a liposome or a carrier virus. Unlike most viruses, this virus does nothing other than carry the bacteria.

Germline gene therapy is the other version of gene therapy, where someone modifies the genes inside the egg or sperm cells to treat fertility problems. Scientists first tried to use this form of gene therapy by injecting DNA into the mouse eggs. Scientists learned from the experiment that certain genetic defects could be corrected before the living being was born. However, germline gene therapy could create new mutations, or disrupt normal genes from functioning.

Although gene therapy is risky, progress is still being made towards its safety and reliability. Gene therapy can treat diseases through somatic and germline therapy. Gene therapy can be used for for monogenic diseases, polygenic diseases, and infectious diseases. The three main approaches to gene therapy are replacing a mutated gene with a healthy copy of the gene, inactivating a mutated gene that is not functioning, and bringing in a new gene to help fight a disease. This is how gene therapy can be used to treat diseases.

***How does somatic cell cloning work? Is it safe?**
Maryam Khan

Cloning has stimulated the minds of scientists for a very long time. It can be therapeutic or reproductive depending on the goal trying to be achieved. Somatic cell nuclear transfer (SCNT) is the method used to reach these goals. For one, Dolly the sheep is an example of a successful outcome of reproductive cloning. Although cloning has been achieved previously, there are still many factors and problems that inhibit scientists to move from cloning sheep to humans.

One of the two types of cloning is reproductive cloning. In this type of cloning, a new organism is created, and is genetically identical to its donor. One method to achieve reproductive cloning is called somatic cell nuclear transfer. A somatic cell is any biological cell that is not a reproductive cell. The procedure starts out with the removal of the nucleus of a somatic cell. This cell comes from the donor animal. The somatic cell, now with without a nucleus is called an enucleated cell. An egg cell is then taken and becomes enucleated as well. The nucleus of the somatic cell is then inserted into the enucleated egg cell. This cell is now called the zygote. The zygote is then stimulated and put inside a surrogate where the egg develops into a fetus. The surrogate will give birth to an identical copy of the donor organism. Although this may seem as a relatively easy concept, many factors are at play.

Therapeutic cloning is another method of cloning that helps patients with degenerated cells. Replacement of regenerative tissue can happen using a zygote. This treatment does not involve the DNA of another human which eliminates the risk of host cell rejection. The process of therapeutic cloning starts with the zygote multiplying into a blastocyst. The blastocyst has 2 layers and inside and outer. The inner has a layer of stem cells which are able to become cells that can replace the degenerated cells. The blastocyst is broken open and is grown to yield stem cells. These then are used to replace damaged cells.

The risks and benefits of cloning have been compared to decide if it should be experimented on humans. The use of SCNT can expose the fetus to risks that can cause complications during birth and later in the child's life. When Dolly the Sheep, the first mammal cloned using an adult somatic cell, was created through reproductive cloning, scientists had been researching and experimenting for a long time and were provided with a nearly unlimited supply of cells and eggs. With humans however, this unlimited supply is not available. Human oocytes, immature ova, also tend to be more fragile and susceptible to failure if not under the correct conditions. As well as reproductive cloning, therapeutic cloning bears its own risks. Adult cells are limited but essential for therapeutic cloning and only few stem cells are usable. Cells can also mutate and cause tumors in patients. In addition to its risks, therapeutic cloning is also very expensive.

It is considered among many that the use of cloning is unlawful. Cloning as well as gene editing is considered morally wrong. Ethical issues raised by using somatic cell nuclear transfer to create humans relate to safety problems and questions what it means to be human. The idea of having control over genetic material of a child is inhuman and opposes the way of nature. The risks and benefits of using cloning have been put under careful consideration by consulting both scientists and ethicists. Many people are concerned about how cloning would impede social ethics and provide excuses for those cloned to be treated as objects rather than people.

While cloning may not be as far in the future as we think, there are still many aspects that must be considered. There are many risks as well as moral issues that must be weighed. Past attempts have worked despite many obstacles, but scientists are learning more and more about cloning every day to make it efficient and consistent.

***What is the process used in gene cloning?**
Marggie Yang

We've all heard of gene cloning. You may have heard stories of cloning one's own pet dog. The question is, how? Before diving into the process, let's clear up a few basic and necessary definitions, *genes* and *cloning*. A gene is the basic physical and functional unit of heredity. Cloning is the process of making an exact genetic copy of something.

Now we know what gene and gene cloning are. So, what are the steps?
The following steps will guide us through the twisting mysteries of gene cloning.

Starting with the first step of gene cloning. It is to join different sources of DNA together by cutting and pasting. This process requires a restriction enzyme and DNA ligase, which is an enzyme that joins DNA together. Our next goal is to insert the target gene into a plasmid to form recombinant plasmid, which is a circular piece of DNA.

Next, the second step of gene cloning is to make bacterial plasmids that are cut with the same enzyme. The gene to clone is cut out by restriction enzymes.

Then onto the third step. The gene-sized DNA and cut plasmids are combined in a tube. The DNA are introduced to bacteria using a process called transformation. During transformation, a heat shock or any shock of a sort will be given to transfer the recombinant plasmids into the bacteria. The bacteria will be grown on a petri dish in colonies, though not all colonies contain the proper plasmids that are needed. There are two types of bacteria. Ones that will take up a plasmid, and ones that won't. The ones that won't take up the plasmid are the ones that aren't wanted. Well, how do you select the bacteria that actually acquired the plasmid? There is an antibiotic resistance that is placed into the plasmid. The ones with antibiotic resistance will survive because it has the gene that allows it not to be susceptible to the antibiotics. The ones that don't have antibiotic resistance won't survive. The plasmids that have antibiotic resistance are going to grow on the mixture. The other plasmids that were not able to take up the antibiotic will not grow at all.

The fourth and final step to gene cloning is to bring the successful colony and continue to produce more and more of it. This results in multiple copies of the specific gene.

Now, we have learned the intricate process of gene cloning. In conclusion, the process of gene cloning explains why some things look identical to each other and may lead scientists into the future of cloning extinct animals.

**How does chromosomal inheritance relate to genetic traits in humans?*

Ashley Cong

Have you ever wondered why you looked like your parents? The chromosomal theory of inheritance explains how genes are passed down from our parents, giving us traits that makes us similar to them. Discovered by Sutton and Boveri, the chromosomal theory of inheritance explains that chromosomes, which come in matched pairs and carry genes on a specific location, are essential for all genetic inheritance.

What are chromosomes and how do they affect the way we are? Chromosomes are structures of nucleic acids found in the nucleus that carries genes and genetic information of an organism. There are 23 chromosome pairs in each human cell, 22 of them are autosomes, and each member of the pair resembles the other. The 23rd chromosome pair, the sex chromosomes, are different for males and females. They are passed down by our parents.

There are two types of each gene (the alleles) that may make up a chromosome pair. One is dominant and one is recessive. When a dominant allele is paired with a recessive allele, it overpowers the recessive genes, making the chromosome a heterozygous pair, where dominant alleles tends to be expressed, despite the presence of the recessive alleles. For example, if there is only one copy of dominant allele for brown hair, and another recessive allele for blonde hair, then the person would most likely inherit brown hair. There must be two recessive alleles and no dominant alleles present for one to display blonde hair. Furthermore, with two heterozygous pairs gives four possible arrangements when fertilization happens, resulting in two alternatives since two of the arrangements are the same when both dominant and recessive alleles are involved.

The Punnett square is a widely known diagram used to predict trait. It is a square diagram used to help determine all the possible traits a child could inherit from the parents. It is composed of a table with four or sixteen squares depending on the number of traits being considered with respect to each other. Each resulting square represents one fertilization event of combining male and female gametes. A monohybrid cross with four squares is relevant when predicting one single trait. When predicting hair traits in the Punnett square, for example, if there is a heterozygous cross with two heterozygous genotypes of curly to straight hair, there will be two dominant alleles and two recessive alleles each labeling a box of the Punnett square. Therefore, there will be four possible offspring resulting, but since there were two dominant genes before, the probability of curly hair to straight hair would be 3:1.

In conclusion, the chromosomal theory of inheritance explains how traits are passed down to us from our parents. Chromosomes are nucleic acids that makes up DNA, which comes in pairs carrying specific genes and traits of an organism. Dominant alleles overpower recessive alleles when paired together to form heterozygous pairs. When using a monohybrid cross Punnett square to predict traits for two heterozygous genotypes, the probability for the dominant to recessive trait would be 3:1. In all, the traits we inherit from our parents are carried by alleles on chromosomes, which can be predicted using the Punnett square.

***What are the genetics of Huntington's Disease?**
Brody Rolston

Huntington's Disease is a rare neurodegenerative disease that is responsible for many physical and mental problems. Huntington's disease was originally named Huntington's Chorea after George Huntington, the author of the disease's first main report in 1872. Its name has since been changed to Huntington's Disease as chorea (movement disorders) is not the only symptom. While symptoms of Huntington's have affected all racial groups, they are most common in people with a North European heritage. Its prevalence is roughly estimated to be about 7-10 per 100,000 people. Even though this number may seem small, the deadliness of Huntington's serves as a clear reason for its infamy in modern press, media, and in society in general. Thus, researchers have worked endlessly to analyze the cause, effects, and possible treatments of this horrible disease.

Huntington's is caused by the repeat of specific nucleotides, the building blocks of DNA, that result in brain damage. Specifically, it originates in chromosome 4, where a gene called Huntingtin exists. In it, the nucleotides cytosine, adenine, and guanine (C.A.G), are continually repeated in that specific order to code for the amino acid glutamine. In a normal person, these C.A.G repeats can occur anywhere from 10 to 35 times. However, people with Huntington's have over 36 repeats. While it remains unclear as to how this specifically contributes to the symptoms of Huntington's disease, it is known that the excess C.A.G repeats result in excess glutamine. This glutamine most likely clusters together in the Caudate and Putamen, parts of the brain usually associated with inhibiting movement, causing its cells to die.

Even though Huntington's effect on a single person is deadly, Huntington's can easily spread to an offspring through sexual reproduction. When a fetus is developing, its DNA continuously replicates. However, when Huntingtin is being replicated, DNA polymerase can lose track of which C.A.G repeat is being copied and accidently insert additional C, A, G's in a process known as repeat expansion. Thus, the child of someone with Huntington's disease can inherit more repeats than their parent. While the typical age of onset is about 40 years of age, increasing the number of repeats makes this age much smaller. This phenomenon is called anticipation, where Huntington's families experience earlier symptom onset with each passing generation. The implicated symptoms from repeat expansion and Huntington's disease in general are extremely serious. Huntington is a dominant condition, so only one allele is needed to produce the condition. The problem is that the onset of symptoms begins after the afflicted person has had children. There exists, then, a 50% chance that the child will have the disease.

Symptoms of Huntington's disease can be described as progressive disturbances to the central nervous system. The results of which include movement, cognitive and mood-based problems. This is because Putamen and Caudate cells work to prevent the body from moving randomly. When these cells die from the excess glutamine, multiple movement problem ensue. Chorea is just one example, where the body makes seemingly random jerky dance-like movements that can only be suppressed with sleep. In addition, the death of these cells can also lead to things such as depression, dementia and frequent personality changes. As time goes on, these symptoms only get worse due to the rapidly depleting number of Caudate and Putamen cells. Eventually, these symptoms amount to death in about 10-20 years after these symptoms begin, usually by the hands of suicide and aspiration pneumonia. This sad truth is made even more terrifying by the fact that a true cure does not yet exist.

As seen prior, Huntington's Disease is a deadly disease that causes movement and cognitive problems. Due to the excessive repeats of the nucleotides, cytosine, adenine and guanine, an abnormal amount of glutamine is created. This clogs the Putamen and Caudate, parts of the brain responsible for inhibiting movement, which causes their cells to die. The results of which causes progressively worse movement and cognitive problems, such as dementia and personality changes. This is made more terrifying by the relative ease in which Huntington's Disease can be passed on and from absence of a true cure. In conclusion, Huntington's is a deadly and terrifying disease that has built its infamous reputation for a reason.

What are mutations, what causes them, and what can prevent them?
Eric Hu

Mutations are rare occurrences that are diverse in type and cause. In general, mutations are defects in DNA or RNA that are caused by the substitution, insertion, or deletion of a base or base pair. What is even more interesting is that evolution has provided us with a defense mechanism to prevent and destroy faulty RNA. Mutations are very diverse and intriguing phenomena that are deeply intertwined with the human body.

Substitution mutations are simply mutations where one base is substituted in for another. For example, the codon ACG would normally result in the amino acid Threonine. However, if the middle "C" was substituted with a "T", then the amino acid produced would be Methionine. In this specific example, the resulting amino acid changed. When that happens, it is called a missense mutation. In addition, there are
nonsense mutations where what would be an amino acid codon is changed into a stop codon. Lastly, there are also silent mutations, where the amino acid is not affected at all. It is important to keep in mind that the three classifications of mutations do not just apply to substitution mutations, but also to a potentially more dangerous type of mutation called frameshift mutations.

Frameshift mutations are another type of mutation that can be very dangerous because of how they can mess up the rest of the strand. Frameshift mutations are insertions or deletions of nucleotides in a codon. Unlike substitution mutations where typically a single nucleotide was substituted and only a single amino acid changed, frameshift mutations affect the amino acid where the mutation occurs and every other codon behind it. If a strand of RNA reads TCTGCTAACG (Serine, Alanine, Asparagine, and an extra Guanine), and the second "T" is deleted, the strand would then read TCGCTAACG (Serine, Leucine, and Threonine). A simple deletion of one nucleotide can mess up the whole chain which is precisely why frameshift mutations can be so dangerous. Now, with the types of mutations covered, it is time to now see what causes mutations in the first place.

The causes of mutations can be divided into two main categories, hereditary and somatic. Hereditary mutations are mutations that were passed down from a parent. This means that either the sperm, egg, or both had a mutation present and as the embryo fertilized, the embryo acquired the mutation as well. Somatic mutations are a bit different as one acquires these mutations after birth. Somatic mutations are typically caused by errors in DNA copying when the cell undergoes mitosis or by ultraviolet radiation and other environmental factors.

Luckily for humans, there is a natural defense mechanism that helps to prevent most mutations in the mRNA before they are used to make proteins at the ribosome. mRNA stands for messenger RNA and this specific type of RNA is translated from DNA and is then transcribed at the ribosome to make codon chains and proteins. A process called nonsense mediated decay (NMD) is operated by a group of special proteins that looks at faulty RNA and decays them. The main protein here is the UPF1 protein and it processes all RNA and decays the ones with faults that it finds. Think of it as a quality control manager at the end of the production line at a factory. Considering how people can still have mutations, these proteins are clearly imperfect, however, they still offer a large line of defense against many potentially dangerous mutations.

In short, mutations are randomly occurring anomalies in RNA and DNA that are very diverse. From substitution mutations to the more dangerous frameshift mutations, these mutations can be caused through either hereditary means, or somatic means. Lastly, NMD can prevent many mutations from happening in the first place. Overall, mutations are a very deep and complex subject, yet they are still connected to us by defining who we are on a molecular level.

**How do environmental and genetic factors affect personality – (twin studies)*

Julia Wang

Identical twins are crucial to understanding the environmental and genetic impact on certain traits, illnesses, and disorders. Twin studies can be used to determine the extent of influence environmental factors have on personality due to the genetic aspect being constant. Furthermore, it is this unique factor that has allowed scientists to better understand our very nature. This section delves deeper into twin studies, epigenetics, the Nature vs. Nurture debate, and how scientists can determine the genetic influence of a specific trait.

Identical twin studies are studies that are conducted on identical twins in an attempt to figure out the extent of influence genetic and environmental factors have on traits, disorders, mental illness, and phenotypes. The two main types of twins are monozygotic twins and dizygotic twins, but only monozygotic twins are used in twin studies. Monozygotic twins, more commonly known as identical twins, are twins that have split from the same zygote (fertilized egg), making these twins completely genetically identical in every cell. Identical twins are born with the same genetic code, but as they age their environment will be different and they will experience different things. This aspect makes identical twins the perfect subjects for determining the extent of influence genetic and environmental factors have on personality.

Epigenetics is the study of changes in gene expression rather than the alteration of the genetic code, studying changes in phenotypes without changes in genotypes. In order to understand what epigenetics is, an understanding of DNA and genetics is needed. Deoxyribonucleic acid (DNA) is made up of four main nucleotide bases. DNA lies in every single cell in an organism's body and the order of these bases contain the instructions needed for the organism to survive. Epigenetics affects genes by determining what genes are expressed or repressed in a cell so that it may fulfill its specific purpose. For example, while a muscle cell and a hair cell have the same DNA, they both have very different functions. Environment is also a major part of epigenetics, as certain parts of an environment can help turn on or turn off gene expression as well. In the case of twin studies, since the genome of identical twins happens to be the same, any differences between them are caused by environmental factors turning certain parts of the genetic code on or off.

The Nature vs. Nurture debate revolves around the argument on whether human behavior is determined by environmental factors in an individual's life or by the genes an individual was born with. The nurture aspect in this debate can include the experiences an individual grew up with in their childhood, including but not limited to: climbing a tree, the culture they were exposed to, relationships they had, etc. The nature aspect refers to genetic code given to them by their parents, your appearance, and personality traits. While neither one of these aspects are the sole reason why someone's personality is the way it is, past debates have argued which facet played a more significant role. For the most part, experts now agree that both nature and nurture have critical roles to play and the way they both interact with each other affects us throughout our entire lifetime. Perhaps instead of "Nature vs Nurture", the way we should be examining this topic is "Nature AND Nurture".

The heritability estimate is a mathematical formula that can give scientists an estimate on how certain genes can impact an individual's personality in a specific environment. Heritability is a concept of statistics that depicts how much a specific trait has been influenced by environmental or genetic factors, with the spectrum of heritability varying from a scale of 0 to 1. When a trait has a heritability closer to 0 it means that the trait was influenced more by environmental components rather than genetic. On the opposite end of the spectrum, the closer a trait has a heritability close to 1, the higher influence of genetics and lesser influence of environment. It is important to note that heritability does not give the reason for why any particular individual has a disease. It applies instead to populations. Heritability determines to which degree of variance in a population is determined by genetic factors, and since populations differ from each other in various ways, one population will never have the same heritability as another.

With the close examination of identical twins, scientists have been able to find the answers to topics ranging from mental illness to even space travel. Twin studies allow scientists to determine which traits or personality quirks are influenced by environment, genetics, or in many cases, both. These studies are crucial in understanding how our genetics interact with our environment, and ultimately get us closer to understanding the true nature of humanity.

Is it possible to harness axolotl genes to regrow parts of the human body?

Alysa Shi

Through the study of axolotls, a group of amphibians with incredible regenerative powers, the secret to organ regrowth could be unlocked. The way they heal allows them to regrow limbs up to six times. They also possess the ability to regrow organs. The research of the genes of these special animals have the potential to revolutionize modern medicine.

Humans heal wounds through the formation of scar tissue effectively terminating any chance of regrowth. It starts with hemostasis, which sends blood cells to the wound and begins to form a clot. Then, inflammation occurs. Collagen, blood vessels, veins, and skin are formed and a scar appears.

Axolotls can regenerate because their cells begin to reform the missing limb. The cells nearby will revert to an embryonic state and come together to make a lump known as the blastema. These cells will then differentiate themselves again and begin to take the shape of the missing limb. Their immune system plays a very big role in the healing process of the axolotl. White blood cells called macrophages control inflammation that would otherwise block the regrowth of the new appendage. Without them, regeneration cannot successfully occur. Their fibroblasts, or connective tissues, provide the location of the wound and instruct their body on what piece needs to be formed. Research has found that retinoic acid is able to rewrite the map that fibroblasts provide, and as a result, scientists have been able to form things like extra-long limbs. One of the most impressive things is that they can regrow function parts of their ovaries, lungs, heart, and even their brain. However, though they are starting to figure out limb regrowth, the regrowth of these internal organs is still quite a mystery to scientists.

Deciphering the genes of the axolotl could lead to a revolutionary change in the way we treat limb and organ loss. The genome was recently completed by an international team of scientists. Their findings were published in a scientific journal and allows other scientists to conduct more specific research about parts of the axolotl healing processes. It is also believed that based on the current research, the thyroid is a significant piece as well. It could help maintain the axolotl's neoteny, or its youthful qualities that help it to regenerate. The best part is that the research above is only just the beginning of a new medical revolution.

The close examination of the axolotl genome and healing process could lead to the ability to regrow parts of the human body. By looking at the similarities and differences, scientists may be able to find a way to allow people to reclaim their lost limbs.

Nervous system

**How are our cells masters of electrostatics?*

Ross Gao

Evolution has seen organisms competing to utilize even the most unconventional energy sources, including that in the realm of electrostatics. Down at a cellular level, vast amounts of energy are stored in the movement of electrons and charged ions, which is in turn governed by the simple laws of electrostatics- opposite charges attract, same charges repel. This leads to an essential principle of cellular mechanics, that ions move down their charge gradient, away from higher concentrations of its charge. The simplicity of this underlying principle facilitates cells' harnessing of the energy, which is then readily employed to drive various cellular activities.

To elucidate further, let a few examples be examined. Photosynthesis, an extremely important biological process which accounts for the creation of all our food, cannot function without electrostatics. Arguably, the most important part of this mechanism is the synthesis of ATP. The plant of ATP production is the ATP synthase, located at the end of a complicated chain known as the electron transport system in the Thylakoid membrane. An ATP synthase takes an Adenosine Diphosphate (ADP) and a phosphate group at a time and puts them together, with a movement similar to that of a rotor. To turn this rotor, a constant inflow of protons (Hydrogen ions) is required, from the outside of the thylakoid to the inside. This, however, can only occur spontaneously with a concentration gradient across the membrane- that the concentration of protons inside the thylakoid is greater than the outside. The task of creating this gradient rests on the electron transport system. Water is broken down at the beginning of the ETS, providing electrons, which is then passed down the chain. Each time an electron sways across the membrane, two protons is taken inside withal and deposited; this process repeats three times throughout the chain, and thus comes a steep gradient of protons. The only channel for protons to move down their gradient- another way of saying to escape- is through the synthase, which awaits their electrostatic energy to create ATP. As such, the thylakoid utilizes the electrostatic forces between charges in a process as complicated as photosynthesis.

An important phenomena of animal cells is resting membrane potential, a voltage across the cellular membrane. Generally, the inside of an animal cell is more “negatively charged” compared to the outside. This is caused by a difference in ion concentration; and just as in the case of plants, it is constructed and maintained to provide energy for cellular activities as ions move down their concentration gradient. This is not to suggest, however, that the two scenarios have no differences. While there is only one type of ion involved in photosynthesis- hydrogen ions- there are three main types of ions involved: potassium, chlorine and sodium; potassium is more concentrated inside the cell, while the latter two are more concentrated outside the cell. This is accomplished through ion pumps, most notably the sodium-potassium pump which pumps three sodium ions out in exchange of pumping two potassium ions in. In addition, it can be observed that animal cells swell and even rupture if put into water: such a phenomenon is partly the product of electrostatic regulations. The swelling of animal cells is the product of osmosis, the movement of water across the cell membrane. While in the case of photosynthesis hydrogen ions are expected to move down their gradient constantly (or at least when there is sunlight), animal cells only open ionic channels in times of need, whereby the movement of ions is greatly restricted. When the solute cannot move, the solvent moves, leading to the phenomenon we observe.

One of the ways the gradient can be exploited in animal cells is the process of nervous signal transmission. In this process, excited nerve cells open some of their sodium channels, allowing sodium ions to flood in. If the stimulus is strong enough and enough sodium ions enter the cell, a threshold reaction is triggered, leading to even more channels being opened. This sudden inflow of sodium ions, centered at a small area of the nerve cell, gets transmitted down the axon of the cell as sodium ions repel each other. Once the signal is past a region, this region of the cell immediately restores the resting membrane potential through sodium-potassium pumps, readying itself for a new signal. A normal nerve cell can complete this task at an astonishing speed, allowing it to transmit hundreds of signals a second.

From plants to animals, the organisms of the biosphere are all masters of electrostatics. With various mechanisms, mainly channels and pumps, cells create and in turn utilize concentration gradients of a variety of ions (or electrons). The functions of these mechanisms are essential to some of the paramount biochemical processes. Indeed, it is in fact quite literal when people say that it is time to charge our cells to work.

What is memory?
Shakthi Velmurugan

The process of memorizing can be split into four sections: encoding, consolidation, storing, and retrieval. These functions filter out the wide variety of information, helping people focus on important things and stay sane.

Encoding allows for an item to be converted into a construct that can be recalled later. It starts with perception (through the senses), which requires attention. This attention is regulated by the thalamus and the frontal lobe, regions of the brain in which neurons fire more frequently. As the neurons fire again and again, the pathway is engraved into our memory, and we remember it. The amygdala is responsible for processing sensitive memories, and we tend to remember these better than regular memories. The sensations are decoded in the sensory areas of the cortex, and then combined in the hippocampus as one experience, which analyzes the inputs and decides if they will go to long-term memory. The memory is compared with previous experiences to determine where it will be stored. There are four types of encoding, namely acoustic encoding, which encodes sound, visual encoding, which encodes images, tactile encoding, the encoding of how something feels, and semantic encoding, encoding sensory input that has a meaning or that can be applied in context, rather than being obtained from a sense. During visual encoding, the amygdala, which processes emotional information and is most noted for its fear conditioning, plays a vital role. It is believed that long-term memory is more reliant on semantic encoding. Human memory is associative. This means that people remember things better if it can be connected to an existing memory. If the memory is more meaningful to the person, the person will encode and consolidate it. Hard to understand information will not be remembered as effectively or may be distorted.

Consolidation is the process through which memory is stabilized after it is first received. This is done through potentiation, which is the process through which a certain pattern of neurons firing make it more likely that that neuron pattern will fire (occur in the same order) in the future together. Eventually, they become sensitized to each other, and they will always fire together when it is recalled. The brain can rewire new connections depending on new experiences that accumulate. This eventually creates a neural network. One of the most important neurochemical foundations of memory is synaptic plasticity or neural plasticity. This is the ability of the synapse (connection between two neurons) to change in strength. Multiple memories can be encoded in a neural network, by different patterns of synaptic connections. Sleep, spaced repetition, reflexion or deliberate recalling of memories are some voluntary and involuntary activities that help consolidation. Re-consolidation, however, can change the initial memory. As the memory is reactivated, the memory can become associated with new emotions or environmental connections, leading to a change in the memory.

Storage is the passive process of absorbing information in the brain. This includes short term memory, sensory memory, and long-term memory. Long term memories are spread out over the cortex and are not clumped in a bunch in one region of the brain. If one memory is forgotten, then the memory can still be retrieved through different pathways.

During retrieval, people can recall information that was previously stored in the brain. This is called remembering. There is no solid distinction between thinking and remembering. The main difference between the actual experience and the memory is the awareness. Memories are not frozen in time, and so can be changed or altered depending on the current situation. They can also be incorporated into other memories. When a memory is recalled, it is called from the long-term memory to the short term or working memory. From here, it can be modified and strengthened, and then it is sent back to the long-term memory. Most of what we remember is linked to questions or cues and is called direct memory. Other memories are stored using hierarchical inference, which is when certain memories are linked to subsets of different groups. Also, the brain is quickly able to determine that there may not be a point to trying to find the answer to a certain question. There are two methods of accessing the memory. One is recognition, which associates events and objects with those that have been previously encountered and compares the information with the memory. This is simply deciding if an object has been encountered before or not. The next one is recall, which involves memorizing facts, events, and objects that are present and recalling them when they are not present. While shallow processing creates a weak memory, deep processing can create a strong memory. Since recalling is a more-or-less automatic process, distraction during recall will not affect the accuracy of the memory, but it will slow down the process. However, distraction when a memory is being encoded can affect the memory.

Thus, memories are created by the four different processes of encoding, consolidation, storage, and retrieval.

*How does the brain affect what one sees?

Shota Gen

Many people believe that sight is reliant on the eyes. However, what one sees is not primarily affected by the eyes but by the brain. The eyes are the source of the information for a person's sight, but the brain is where the image of what they actually see comes from.

All of what one perceives is affected by the cerebral cortex, the outer layer of the brain, that is responsible for processing the information of one's senses. The cerebral cortex is divided into four lobes by sulci and gyri. Gyri (Singular: Gyrus) are the bumps and sulci (Singular: Sulcus) are the grooves on your brain that contribute to the brain having a greater surface area within the volume of your skull. The four lobes have their own association areas that is central to someone's sensual perception. They are the frontal lobe, parietal lobe, temporal lobe, and occipital lobe. The visual association area, used for perceiving visual data, is located in the occipital lobe at the back of the brain.

How the brain processes the sensory information is split into two ways: bottom-up processing and top-down processing. Bottom-up processing is when the object that one looks at is the primary source for their perception. This processing is data-driven in the fact that one does not use previous knowledge to discern an object, but only uses what they see. However, when the object is made up of smaller parts the brain uses previous knowledge to recognize those smaller parts and tries to comprehend the bigger picture. Top-down processing uses previous knowledge to identify an object even though it may be different in some way. As an example, one might see an incomplete drawing of a circle. The brain, however, recognizes the incomplete illustration as a circle using previous knowledge and assumes what it is.

There are many different types of perception that contribute to the overall perception of one's sight. There is form perception that recognizes the visual aspect of the objects someone sees, which is divided into three sections. The first is proximity, grouping nearby figures together. The second is continuity, perceiving continuous patterns and not broken ones. The third is closure, filling in gaps to create something one recognizes. There is also depth perception, which recognizes the 3D characteristics of an image although it may be 2D. The last is motion perception, which infers speed, direction, and distance using motion. For example, one can tell an object is retreating when the size of the object is shrinking and vice versa.

There are different factors that contribute to one's overall perception of their sight. The eyes only feed the brain information of one's vision, and it is up to the brain on how one sees that image.

**What is the limbic system and what is its function?*

Tejas Shivaraman

The Limbic system has been controlling our every move ever since we first stepped onto this Earth. Every emotion that drives us, every memory we recall, has been driven by the Limbic system. The Limbic System includes the Amygdala, Hypothalamus, Thalamus, Hippocampus, Frontal Lobe, Olfactory Bulb, Prefrontal Cortex and the Cingulate Gyrus. You can find the parts of the Limbic System inside the brain beneath the cerebral cortex but above the brainstem. The Limbic System has two key functions within the human body: Emotions and memories.

The amygdala controls emotional responses such as love and fear. If the amygdala was damaged or overstimulated, it could result in abnormal emotional reactions or excessive reactions, respectively. The hippocampus sends information to the amygdala. When it comes across a memory with emotional ties, it interacts with the amygdala. The prefrontal cortex correlates with making decisions while in an emotional state. Like the hippocampus, the hypothalamus feeds information to the amygdala. It also acts as a regulator of emotion. Finally, the cingulate gyrus provides a pathway between the thalamus and hippocampus. Its main function is to alert the rest of the brain that something emotionally significant is happening.

As stated before, different parts of the limbic system interact with different functions of the system, and in the case of the memories, the two key players are the hippocampus and the amygdala. The hippocampus is essentially the memory hub of the brain. This is where our memories are formed and then categorized to be put out into long-term storage. It also connects memories with various senses such as touch or smell. The olfactory bulb, for example, relates smell to memories. It transports smell from the nose to the hippocampus and as a result, we relate this smell to a specific memory. The other key player, the amygdala, has a different approach to memories by linking them to emotions. This connection causes memories that have stronger emotions to be stored more strongly. When people think of the amygdala, they think of fear. It is the reason humans are afraid of things outside their control. When humans have memories associated with fear, the amygdala kicks into high gear. These memories only need a few repetitions to be put in our brain compared to the multiple repetitions needed with normal memories. The amygdala is also heavily involved with arousal and stimulation, however, there is still not much known about the role of the amygdala relating to sexual stimulation and arousal.

The limbic system provides us the ability to love one another. It has also enabled us to recall events that we hold dear.

***What are the characteristics of a Down syndrome person?**
Sydney Moore

Down syndrome is the most common chromosome abnormality affecting mankind. Someone with this abnormality is usually diagnosed when they are born; it is not something they could possibly develop as they travel through life. Someone affected by Down syndrome can face many challenges, such as difficulty staying focused for long periods of time or grasping concepts as fast as other people without the condition.

Down syndrome is a chromosomal deformity that 1 in every 700 babies is born with. Each cell in the body contains a nucleus which carries genes. Genes hold all the genetic material that someone inherited from parents. Chromosomes are structures that carry the genes which are all grouped together. Typically, a person will have 23 pair of chromosomes, one of each pair being donated by one parent. In some cases, someone may have an extra copy of chromosome number 21, and this is when Down syndrome occurs.

The characteristics of the brain in someone affected by Down syndrome are significantly different than a normal brain. The brain and therefore head of someone with Down syndrome is significantly smaller than someone without. Brains of young adults with Down syndrome do not show any significant abnormalities, except for a subnormal weight and a simple convolutional pattern. Convolutional patterns are one of the unusual ridges on the surface of the brain. In humans with Down syndrome over the age of 35, there is a development of more noticeable abnormalities of the brain. One abnormality is senile plaques, extracellular deposits of amyloid beta in the grey matter of the brain. Amyloid beta is an amyloid that is taken from a larger precursor protein. Another abnormality is granulovacuolar neuronal degeneration which is the degeneration of hippocampal brain cells in people over the age of 35. Hippocampal brain cells are the cells responsible for learning information, storing long-term memories, and regulation of emotions.

A brain affected by Down syndrome changes the physical appearance of the person. There are many common noticeable physical traits with people affected by Down syndrome. These include low muscle tone, small stature, an upward slant to the eyes, and a single, deep crease across the center of the palm. The differences in the way people look are a common way to tell if they are affected by Down syndrome. There are cases, however, where there are no physical abnormalities at all, and the person affected by Down syndrome does not look different in any way.

People with Down Syndrome are affected in many ways. Down syndrome is a common chromosomal abnormality, and there are more deficiencies developed in the brain in adults who are over the age of 35. This is how the body of someone with Down syndrome is affected.

**How do sensory neurons convey signals from sensory receptors to the central nervous system?*

Su Kyung Lee

Sensory receptors in ears, eyes, nose, tongue, and skin detect changes in the environment and turn it into electrical impulses. The nervous system is divided into the central nervous system and peripheral nervous system. All sensory messages are brought to the central nervous system, which is responsible for regulating and controlling one's senses. The peripheral nervous system includes the cranial nerves to the nerves that connect the spinal cord to the body.

Sensory neurons begin in the periphery where all the stimuli are generated by a network of nerves. When sensory receptors in the skin are exposed to any stimulation, they automatically transmit an impulse which ultimately reaches the brain.

Transduction takes place in the peripheral sensory when sensory cells translate chemical, electromagnetic, mechanical stimuli into action potentials. The sensory neurons involved in vision are visual sensory neurons. These neurons contain rod and cone cells in the retina that convert the energy of light signals to electrical impulses that travel to the brain. In the auditory system, hair cells in the inner ear convert vibrations of sound into electrical energy. The nerve fiber carries the information of the stimulus towards the central nervous system. Depolarization, when the cell experiences a change in electric charge distribution, is occurred by the process of transduction.

In sensory neurons, dendrites, short extensions of nerve cells in the periphery pick up stimuli and conduct messages toward the central nervous system. The impulse is sent up the spinal cord and through the brainstem to the thalamus, a center for processing sensory information located deep in the brain. The impulse crosses a neural connection in the thalamus to nerve fibers that convey the impulse to the sensory cortex of the cerebrum (the area that receives and translates information from sensory receptors).

The cell body of sensory neurons are near the spinal cord and are gathered near the dorsal root ganglia. Stimulus is detected by a receptor which sends electrical impulse message along with a sensory neuron to the central nervous system. Dendrite carries the signals toward the part of the cell that encloses the nucleus. The central nervous system then conveys the message by motor neurons to effectors which provide a response.

Next, axons on terminals on synaptic knobs (in spinal cord or brain) will allow the message to continue along a second neuron. Axons are surrounded by myelin sheath made of Schwann cells. Electrical signal moves down the axon and reaches synapse, the gap between the neurons. When chemicals reach next neuron's dendrites, they stimulate that neuron to begin sending an electrical signal down its axon, and this process repeats.

The process in which sensory neurons reach the central nervous system is complex, requiring sensory neurons carry electrical signals (impulses) from receptors to the CNS. Through this course, we, humans can recognize temperature, pressure, light and pain.

What are some brain tumors in different sections of the brain and how do they affect the function of that specific part of the brain?

Ruhaan Doshi

Before learning about different types brain tumors, one must understand what a brain tumor is. A brain tumor, or any kind of tumor for that matter, is an uncontrolled mass of abnormally growing cells. There are many different types of brain tumors, each of which can be classified based on where, or how, the tumor cells are oriented. Brain tumors can also be can be benign (non-cancerous) or malignant (cancerous). Tumors are categorized into four different grades: grades I and II meaning that the tumor is slow growing and generally less fatal, and grades III and IV meaning that the tumor is fast-growing, dangerous and often, malignant.

A pineoblastoma, commonly found in kids, is classified as a rare and often fatal tumor. A pineoblastoma, as the name suggests, is one of many different types of tumors of the pineal gland. The pineal gland is a small endocrine organ located deep inside the center of the brain is responsible for the secretion of melatonin; melatonin is the hormone that "regulates" sleep. This tumor can cause excess sleepiness as it can cause the pineal gland to produce excess melatonin. A pineoblastoma arises from embryonal cells. This specific tumor is commonly found at the base of the cerebral hemisphere and is usually located near the hypothalamus. As a pineoblastoma evolves and gets larger, it blocks the cerebrospinal fluid (CSF) and prevents it from draining from the brain; the CSF flows through the third ventricle, thus a blockage can cause an unnecessary build-up of it. This, in turn, leads to a build-up of pressure within the skull— a medical condition called hydrocephalus. The hydrocephalus can cause minor vision loss due to too much pressure in the skull.

An Optic Glioma is a tumor that can form on the optic nerve. The optic nerve is situated in the forebrain just behind the eyeballs. A glioma like this one is generally found in children around the age of six and is generally benign. The optic glioma generally forms in the optic chiasm— this is where the left and right optic nerves cross just behind the eyeballs. The tumor presses against the optic nerves and can cause vision impairment, and in some cases: blindness. This tumor, only in the most extreme cases, can grow so large that the eyeball bulges out of its socket.

A glioblastoma multiforme (GBM) is generally located in the cerebral hemispheres in the brain. They are developed by clusters of star shaped glial cells (astrocytes and oligodendrocytes: these are the different types of cells) in the brain. A GBM is one of most common, yet fatal malignant brain tumors that exists today. A glioblastoma multiforme is classified as a grade IV due to its characteristics. These tumors are can grow in many different parts of the brain but are more commonly in the frontal lobe. A brain tumor here can cause intracranial hypertension if the tumor gets too big. This will inhibit the frontal lobe to carry out its duties including emotions, and other cognitive functions. A GBM can also cause swelling of the brain. If it gets too large it can cause headaches, nausea, etc.

There is a wide array of brain tumors that can spawn in many different parts of the brain. Brain tumors are generally seen as life-threatening, but some are relatively harmless to a certain extent. As a brain tumor gets larger and grows faster, the part of the brain that the tumor is in takes a hit and in some extreme cases even shuts down.

***What are the causes of Parkinson's disease, how does it affect the pathways of the nervous system, and what treatments are there to help control the symptoms?**
Sean Barcino

Parkinson's disease affects 10 million people worldwide and is considered one of the "most common progressive, neurological disorder in the US", according to the American Parkinson Disease Association. Parkinson's disease is caused by the loss of dopamine producing neurons in the substantia nigra and can affect the pathways of the central nervous system, but there are treatments being used to reduce the symptoms of the disease.

Parkinson's disease is believed to be caused by abnormal amounts or misfolding of alpha-synuclein, a protein, which causes the loss of neurotransmitters (specifically dopamine) in a small section of your brain called the substantia nigra. Alpha-synuclein is originally made with the help of a gene called SNCA gene. SNCA genes exist mainly in the brain, but are also found in the heart, muscles, and other tissues. The formation of Parkinson's disease is believed to occur when there are mutations of SNCA gene. For example, the formation of amino acids within the SNCA gene may change, causing a-synuclein to be misfolded. Or, SNCA gene may become duplicated or triplicated, creating an abnormal amount of a-synuclein. Whatever the case may be, the misfolded or abnormal amounts of a-synuclein may group together to form lewy bodies, which can kill dopamine producing neurons in the substantia nigra. This causes Parkinson's disease, which can affect specific pathways in the brain.

Parkinson's disease disrupts the direct and indirect pathways of the brain. The direct and indirect pathways are key frameworks of the central nervous system and are responsible for movement / non-movement. The direct pathway is responsible for movement and the indirect pathway is responsible for non-movement. The direct pathway begins with the thalamus, a part of the brain, which sends excitatory messages to the motor cortex, making the motor cortex more active and causing it to send messages to move our muscles. In the indirect pathway, the globus pallidus (internal) inhibits the thalamus. As a result, the motor cortex will not get excited and doesn't send messages to the muscles. But in the direct path way, the substantia nigra inhibits the globus pallidus (internal) from inhibiting the thalamus, so the thalamus can excite motor cortex. However, with Parkinson's disease the dopamine producing neurons in the substantia nigra are killed. Therefore, the substantia nigra cannot inhibit the globus pallidus (internal), thus, causing a huge loss of movement in the direct pathway. Because of this loss of movement, people with Parkinson's disease develop a variety of symptoms, such as small tremors, rigidity, bradykinesia, postural instability, etc.

Although there is no cure for Parkinson's disease, there are a variety of treatments to help control the symptoms. Dopamine precursors are a type of drugs that can pass the blood brain barrier and be converted into dopamine. As a result, by increasing the concentration of dopamine, the brain will function more efficiently. An example of a dopamine precursor is levodopa. If Levodopa is taken as a pill, then it is converted to dopamine in the brain. Unfortunately, after 5-10 years, levodopa is broken down by special enzymes (called MOAB) in the brain. In order to prevent this, levodopa is now sent with MOAB inhibitor, which stops the special enzymes from breaking down levodopa. Another treatment for Parkinson's disease is called dopamine agonists, which are drugs that are used to reduce the negative effects of Parkinson's disease by activating dopamine receptors. Dopamine receptors cause more dopamine to be created, increasing brain function. Drugs like pramipexole and ropinirole are dopamine agonists. All in all, there are numerous treatments that one can acquire to reduce the symptoms of Parkinson's disease.

In conclusion, through sophisticated process, dopamine producing neurons are killed by abnormal amounts or misfolding of a-synuclein. Parkinson's disease can affect the direct and indirect pathways of the central nervous systems, causing multiple symptoms. Luckily, treatments like dopamine precursors and dopamine agonists can reduce the symptoms of Parkinson's disease.

***What is schizophrenia?**

Sofia Valbuena

Living a life without the notion of right and wrong may seem unimaginable, yet some individuals with schizophrenia live this way. People with schizophrenia can lack proper judgment due to the voices in their head telling them otherwise. These voices are a conscious form of the disease that talks to the patients and can sometimes drive the patients to harm themselves or others. This mental disease is a very unfortunate one to have, since the patients' lives tend to be greatly affected in a downward spiral, even to the point of suicide. This condition, when the brain attacks the conscious mind, can cause a person to go clinically insane. Schizophrenia can last anywhere from a year to the rest of the patient's life.

Scientists have found that schizophrenia is caused by a genetic defect. Although there is no recognized specific cause, a similarity among all schizophrenic patients is the slight mutation of their genes. This gene is inherited, which means that the child of a patient is more likely to contract the disease rather than the child of a healthy person. The disease can also be influenced by other factors such as prenatal infections or childhood stress. Schizophrenia can also be thrust upon someone due to emotional trauma.

The symptoms of Schizophrenia include the symptoms of other disorders such as PTSD, depression, and bipolar disorder. One can even contract another mental disorder because of it. For an example, a patient diagnosed with schizophrenia can become bipolar due to the stress that schizophrenia places on the brain. The disease's symptoms can be different for each patient, but most experience a loss of judgment, hearing voices, hallucinations, and/or a distorted view of reality. These symptoms can make the patient stressed or panicked, due to the loss of control of the brain, which can lead to other mental disorders.

Treatments of Schizophrenia consist of antipsychotics and anti-tremor. Although the medical treatments can be harsh due to side effects and can be less than effective, prescription drugs can help minimize the symptoms or, if the patient is lucky, can completely diminish the disease from the brain. They can diminish the power the voices have on the patient by clearing their mind and regaining consciousness. As mentioned by Eleanor Longden in her TED Talk, the support of one's loved ones and a change of mindset can also cure schizophrenia. This disease has no specific cure, but different and unexpected methods can heal a patient.

Schizophrenia is a neurological disorder that has taken away the lives of many, both literally and figuratively. Most patients feel lost or disconnected due to losing the ability to control their brain, which can lead them to take their own lives. This disorder is very abstract in the fact that it can be cured in many ways, or not at all; it all depends on the patient and what works for them. Though this disorder is still a mystery to many, it does not make it any less valid. This disorder is very real and can leave devastating effects on its patients.

***What are the causes of depression and how can it be treated?**

Isabel Jacobson

Imagine feeling hopeless, unable to get out of bed to do the simplest of tasks. People with depression must face these feelings every day. Depression is a mood disorder characterized by persistent lethargy and prolonged feelings of hopelessness. Depression is also one of the most common and pervasive mood disorders, affecting almost 20 million Americans and 300 million people worldwide. Although it is more common in women, the disorder affects people of all genders, races, and ages and is said to be the top reason mental help is sought out.

Because depression affects everyone, it is difficult for scientists to find the root cause of the disorder. However, researchers have concluded that there are two contributing factors to heightened risk of depression: environmental factors and biological factors. Environmental factors are considered to be one of the main causes of depression. Often, these factors are the result of stressful life events. For instance, girls who feel the stress of losing a parent at an early age to external causes have an 80% greater risk of developing depression later in life. For boys, this risk is increased almost 400%. Other environmental factors could be substance abuse, difficulty in relationships, financial problems, climate, or adverse childhood experiences. In addition, diseases such as cancer, thyroid disease, and chronic pain are all linked to greater risk of depression.

Furthermore, environmental factors often collaborate with biological factors to increase the risk of depression. Depression is most likely caused by a chemical imbalance in the brain. Many scientists have found that in a depressed person, the amount of serotonin in their brain is lacking. Serotonin is a neurotransmitter that regulates mood, social behavior, appetite, sleep, and memory. It is made from the amino acid tryptophan, which is an essential amino commonly obtained from cheese, nuts, and red meat. In addition, the amount of norepinephrine in a depressed person's brain is significantly lower than normal. Norepinephrine is a chemical released in response to stress and is linked to arousal and focus. Overall, both biological and environmental factors work together to magnify the risk of depression in a person.

Depression can be diagnosed in a variety of ways, but since it is not a visible illness, it takes a bit of observation. First, a doctor, such as a psychologist, psychiatrist, or therapist, will observe the patient for two weeks looking for five or more symptoms of depression. Symptoms include continual sadness, disinterest in enjoyable activities, fluctuating weight and sleep patterns, loss of energy, and suicidal thoughts and actions. In addition, some doctors use brain scans to determine if the patient has depression. Brain scans of someone with depression show less activity in normally active parts of the brain. There are many different diagnoses of depression, including clinical depression, postpartum depression, seasonal depression, and bipolar disorder.

There are many possible treatments for clinical depression. First, a depressed person would visit a therapist for treatment. If their depression is more severe, they may be prescribed antidepressant medication. The most common type of antidepressants is Selective Serotonin Reuptake Inhibitors

(SSRIS). SSRIs stop the reabsorption of serotonin into brain cells, so there is more serotonin at the junctions between brain cells. Of course, this type of medication is based on the hypothesis that low serotonin correlates with depression. Other holistic treatments include meditation and exercise.

In conclusion, depression is a complex mood disorder that is linked to both biological and environmental factors. Depression is a long-term disease that can have fatal consequences if left untreated. In fact, more than 60% of suicides are linked to depression and similar diseases. Although it takes time and patience, depression is not a permanent disease and most treatment outcomes are positive, improving the lives of millions of people worldwide.

What happened to Phineas Gage and his brain?

Teddy Arida-Moody

Phineas Gage was an average 25-year-old construction worker. In 1848, his life drastically changed. In September of that year, a tamping iron pierced his skull and brain, causing various changes. His job was simple. He'd drill holes into stone, fill the holes with explosive powder, cover it with sand and tamp it down. The incident occurred because Gage lost his focus and tamped down on the explosive powder before his partner poured the sand. The explosive powder exploded immediately, driving the steel bar into his head. The frontal region of his brain was gashed but surprisingly enough, his memory remained intact. Phineas' personality was severely altered though, according to his friends and family. He was one of the best construction workers his company had but after the accident, they had to fire him.

The story of Gage is quite clear, but the real question is, what directly happened to his head. The tamping iron was 3 centimeters thick, 109 centimeters long and a whopping 13 pounds. The iron likely exited his skull 1.5 centimeters from the area of bone loss. According to John Harlow, a scientist who studied Gage, Gage's frontal region was pierced. To be exact, the frontal cortex and the connection between it and the limbic system was damaged badly by the tamping iron. This had a detrimental effect on Gage's personality and behavior.

While the physical damage may seem jarring, the personality changes are far more intricate and interesting. First, a few things should be known about how the emotions travel throughout the brain. The limbic system passes on emotion to the frontal cortex. Then, the frontal cortex assigns priority to these emotions, basically telling your body if they should enact the emotion or not. As stated before, the connection between Phineas Gage's limbic system and frontal cortex was severed by the tamping iron, so personality changes would make sense. The frontal cortex does not participate in prioritizing the emotions once the connection is broken, so the limbic system fires away any emotions it wants. But, the mind-boggling thing about Gage, is that his memory and intelligence were perfectly intact. This got scientists wondering, keep in mind this was the mid- 1800s, about how the brain would possibly store information in different locations, meaning different parts of the brain had different functions. This meant that the part of the brain that affects emotion was damaged but everything else about Gage's mind was secure. Gage was fine after the accident for many years, but these personality changes made him quite awful to be around. He did not abide by typical social cues, thus he was difficult to converse with. He would constantly curse after the accident. According to his close friends: "Gage was no longer Gage".

Gage's story is quite remarkable and has baffled scientists for decades. He passed away in 1861, thirteen years after the accident which was unbelievable for the time and the severity of his injury. Sadly, no autopsy was performed on his body, so scientists can only speculate on how he died. Nevertheless, Harlow, the scientist who studied Gage previously, requested Gage's skull to be examined when he heard about the passing five years later. Lots of tests were performed on the skull, such as x-rays and various scans, and the skull is now kept at a medical museum at Harvard University. Gage will go down as one of the most mind-blowing neuroscience cases of all time, and for good reason.

Evolution

How did the conditions of early Earth allow for life to begin, and what were the first organisms?
Tomás Serna

The Earth is our home, as it is to countless others. When the very first organisms arrived into existence, how was the earth different than it is today? The Earth was formed approximately 4.5 billion years ago, and environmental factors such as the atmosphere, temperature, and natural occurrences were extremely different. The first period on Earth, from 4.5 to 3.8 billion years ago, was named the Hadean Eon, after the Greek god of the underworld Hades. Linking to the image of the underworld, the earth was at a state of molten mass. Over time, the denser elements began sinking to the center of the Earth, creating layers of material but still-remaining molten due to the intense heat and pressure. These layers allowed for volcanoes to form when the Earth started to cool down. The atmosphere was dense because there were large amounts of carbon dioxide. Due to meteorite impacts, it was also filled with large proportions of water vapor. There was no ozone layer, so any organism on the surface of the Earth would be killed instantly due to ultraviolet radiations. The surface temperature was extremely high; around 1,000 degrees Celsius. Very little is known about this period because there is scarcely any physical evidence. Eventually, water from falling meteorites condensed into clouds to provide protection from the sun, and the Earth cooled down much more. This signified the end of the Hadean Eon and the beginning of the Archean Eon. Paired with the release of various gases from volcanoes, the water vapor in the atmosphere condensed into rain. After millions of years, oceans became present on the Earth's surface. The atmosphere was anoxic, meaning that it did not contain oxygen. Therefore, the first organisms had to be able to live without it. Over millions of years, rain washed organic molecules into lakes and ponds containing dissolved minerals, creating an environment where chemical reactions could take place. In these areas, new molecules formed and the first organisms were made.

The first organisms were created 1.04 billion years after the Earth formed. These prokaryotes were single-celled and they lacked a nucleus. The first prokaryotes were anaerobic heterotrophs, meaning that they take energy from others in its environment in order to sustain itself and did not use oxygen. The second prokaryotes were photosynthetic autotrophs, which used sun's energy to produce their own food. Although instead of breaking down water as in modern photosynthesis, these early organisms broke down Hydrogen Sulfide instead because it was easier to break down than water. The next step in the evolution of prokaryotes was cyanobacteria, that break down water in regular photosynthesis. This was an immense advancement because the breakdown of water was able to provide more energy to the organism than the breakdown of hydrogen sulfide. As a result, the population of cyanobacteria increased because they are able to produce more energy. Since cyanobacteria released large amounts of oxygen waste in the air, a new revolution was created where many organisms started using and incorporating it. The next form of prokaryotes, the aerobes, used oxygen directly due to the increased amount in the atmosphere.

Around 3.8 billion years ago, these single-celled prokaryotes started living in the depths of mineral-filled lakes and ponds. The prokaryotes absorbed minerals and ATP in the oceans and eventually rose up closer to the surface. Some started using the sun's light along with water and carbon dioxide in a process called photosynthesis, which allowed them to create high-energy compounds for themselves and let out oxygen as waste. After more prokaryotes developed this crucial technique, the oxygen levels in the atmosphere began rising steadily. However, some other organisms could not survive with the amount of oxygen present in the new atmosphere, so they died off in what is known as the Great Oxygenation Event. The prokaryotes were only the beginning of organic molecules in the world, and their contribution to the addition of water in the atmosphere kickstarted the evolution of more complex organisms and eventually the creation of animals, including humans.

***What is Hardy-Weinberg equilibrium, and how might it be used?**

Jerry Yuan

There are some requisites to learning about the Hardy Weinberg Equilibrium. A human's outside features, or phenotypes, are determined by the random distributions of the alleles of their parents. Alleles are various forms of genes that determine your traits. Often, there are two types of alleles: dominant and recessive alleles, and two alleles combine to determine a trait. If someone has both recessive alleles, then he or she will display the recessive trait. Any other combination of alleles that include one or more dominant alleles will result in a dominant trait. The different combinations of the parent's alleles is called the genotype and the genotypes of an offspring can be represented by the Punnett square. The Punnett square can be used to represent the allele, genotype and phenotype frequencies throughout the population.

Example of Punnett Square:

	P	q
P	pp	pq
q	pq	qq

The Hardy-Weinberg equilibrium is the basis of evolution. Hardy-Weinberg equilibrium means that a population *Does not* have a small population, non-random mating, mutations, gene flow, and adaptation; these five things are also called the five fingers of evolution. First, small population; a Hardy-Weinberg equilibrium requires a large population as described by the Law of Large Numbers, which states that as the number of trials and events becomes larger, the results will approach the theoretical probability of the event. Second, non-random mating: when the probability that two individuals in a population will mate is not the same for all possible pairs of individuals. Humans would be an example of non-random mating because we determine who we want to mate with based on certain characteristics such as charisma. The consequences of non-random mating involve certain phenotypes not being passed on. Third, mutations; mutating is the changing of the nucleotide sequence of a gene, resulting in a variant form that may be transmitted to subsequent generations. Mutations will cause a change in the gene pool and in turn, affecting the outcome of the next generation. Fourth, gene flow; gene flow refers to the transfer of genetic variation from one population to another. An example of gene flow could be the migrating of birds or interbreeding between different species. Finally, adaptation; adaptation refers to a change or the process of change by which an organism or species becomes better suited to its environment. This change to adapt to the environment will cause a change in the gene pool, in turn affecting the results. To conclude, if a population does not have the five fingers of evolution: small population, non-random mating, mutations, gene flow, and adaptation; then it has achieved the Hardy-Weinberg equilibrium and is officially not evolving.

A very essential characteristic of Hardy-Weinberg equilibrium is that the allele frequencies remain the same throughout all generations. This can be represented and explained by the equations: $p + q = 1$ and $pp + 2pq + qq = 1$ or $(p + q)^2 = 1$. In this equation, p represents the allele frequency of the dominant allele in a population, and q represents the allele frequency of the recessive allele in a population. The first generation starts off with an allele frequency of $p + q = 1$. The second generation, assuming this population is at Hardy-Weinberg Equilibrium, will have an equation of $pp + 2pq + qq = 1$. Now, proving that the allele frequency stays the same. If we rewrite the equation of the second generation, we will find that $pp + 2pq + qq = p(p + q) + q(p + q)$. Since we are talking about ratios, we can eliminate the $(p + q)$ and so we are left with $p\cancel{(p + q)} + q\cancel{(p + q)} \rightarrow p + q$.

In addition, there are many ways that we can apply the properties of the Hardy-Weinberg Equilibrium. As previously mentioned, we use the Hardy-Weinberg equilibrium and the five fingers of evolution to define evolution. But we can also use the equation $p + q = pp + 2pq + qq$ to calculate the allele frequencies given the phenotypes of a population. Say the recessive phenotypes of a population were 0.16, this means that $qq = 0.16$, thus we know that $q = 0.4$. Using the equation: $p + q = 1$, we find that $p = 0.6$. The Hardy-Weinberg equilibrium is obviously unrealistic does not tend to occur in nature.

In conclusion, The Hardy-Weinberg Equilibrium is when a population does not have the five fingers of evolution: Small population, Non-random mating, Mutations, Gene flow, and adaptation; and the Hardy-Weinberg Equation is used to calculate evolution under this equilibrium.

What are the Ideas Behind the Theory of Evolution?
Brogan Dougherty

Charles Darwin's Theory of Evolution introduced ideas that still implement today in scientific research and are universally accepted among the scientific community. Some people believe that Darwin was the creator of the concept of evolution. This is false. However, he was the first to publish his research and propose a mechanism for evolution. It is speculated that some of his ideas even came from his father, showing that he had assistance formulating his concepts. The main reason Darwin is so widely recognized today is because he was the first to publish his discoveries. Darwin laid a basic foundation that has helped in the development of more advanced research. For example, Stephen Jay Gould recently proposed that evolution was not necessarily a "ladder of progress", or a continuous improvement of species over time.

Darwin's Theory of Evolution is deep-seated in the concepts of natural selection and adaptation, exploring how the fittest of a species survive and how this impacts evolution. Only the fittest species can adapt to ever-changing conditions. Therefore, if a population is not able to adapt to its surroundings and, for example, effectively avoid predators, offspring will not be produced. Offspring are essential to the longevity of a species and key to adaptations. When an animal is the "runt of the litter," the probability of death increases, while the healthiest ones grow to be stronger and faster. The mother also may tend to favor the strongest of the group, since they have a better chance of continuing the species' survival. This is the basic idea behind natural selection. When coupled with Darwin's ideas of adaptations, a species has a better chance of thriving over time.

Darwin concluded that certain species' adaptations are essential to their survivability in the long term. Many adaptations we see in modern organisms are the result of thousands of years of genetic changes. Adaptations also ensure a species' survival since offspring will be born with traits of the parent to increase their chance of survival. In other words, adaptations are hereditary. Adaptations can vary in severity based on numerous factors, such as climate, availability of food, and how organisms interact within an ecosystem. This is the reason some animals have the ability to live in certain areas because they have genetically adapted to the specific environment. For example, a camel has adapted "humps" on its back to store fat for energy. They have developed this because the arid climate in which they live lacks a consistent source of food. Although Darwin mostly focused on animal species, a flora example may be observed in hot peppers. Certain pepper plants kept getting eaten by various creatures. Those plants that produced small amounts of capsaicin were less likely to be eaten. They had offspring that produced higher levels of capsaicin. Due to the increase of capsaicin, the plant has a burning sensation when eaten. This adaptation helped the pepper plant defend itself because animals were less likely to consume it due to its hot taste. Thus, one can see how natural selection and adaptation were key points in Darwin's theory.

Through the study of different fossil groups, Darwin was able to provide ample evidence to support his Theory of Evolution. Darwin conducted much of his research in the Galapagos Islands where there is a wide variety of species from which he could collect data- in specific, finches. There was also a large variation of types of finches, about 14 species that Darwin studied. The visual characteristics changed a significant amount over quite a short period of time, which meant that they would be easier to study. The reason they evolved so fast is because there was little competition among the birds, so so they could co-exist and eat different foods. The distinctive difference between the current species and fossils were the beak sizes. Some of the finches had to eat different types of food, thus explaining their various beak sizes. They also lived in different parts of the environment. Some lived nearer to the ground and some lived higher up in the trees. The difference in the surrounding environments explained their vastly different sources of food.

These sources ranged from seeds to live creatures such as insect, and various plants. Those with slightly different beak shapes ate slightly different food types. Fossils and live specimens showed the existence of adaptations that were selected for over time. Darwin's findings proved his theory of natural selection and helped further enhance our knowledge of evolution.

An example of someone who has further studied the ideas of evolution is the scientist Stephen Jay Gould. He is most well-known for his idea of Punctuated Equilibrium. Punctuated Equilibrium is the idea that species' evolution depends on short episodes of rapid change. New species can be made quickly which makes the study of their fossils hard to keep up with. The creation of a new species could result from a situation where a species environment rapidly changed, and certain adaptations were selected for, which allowed them to survive. This theory is different that Darwin's theory of Evolution, where the consensus is that species gradually get better over expansive periods of time. The ideas behind evolution were created by Darwin and his research has been essential to the ever-changing idea of evolution.

How does Darwin's theory of natural selection and the theory of genetic drift affect the process of evolution?*

Robert Victor

Humans did not simply appear on earth at one moment in time as they are. Rather, as all organisms, they endured a process of change and adaptation over millions of years–evolution. Humans have evolved from a common ancestor of the apes into the most dominant known species. One might wonder how this evolution took place. There are many proven theories and rationales surrounding the process of evolution and why it occurs as it does. Two such ideas are Darwin's theory of natural selection and the theory of genetic drift. Many different phenomena surrounding evolution can be explained by these two–now accepted–scientific principles. In fact, Darwin's theory and genetic drift each have profound impacts on the evolutionary process.

When discussing the theory of evolution, the name of one man will likely appear more than any other: Charles Darwin. Darwin was an English scientist during the 1800's who made massive contributions to the modern perception of evolution by creating the idea that evolution occurred through a lengthy process where species gained and lost traits dependent on how well they helped organisms survive in a given environment. Darwin called this process natural selection. When visiting the Galapagos islands, he took notice that for each individual island, the breed of finch was slightly different than those on the others. Through his research, Darwin realized that each island's bird shared common ancestors with the rest but had diversified in its evolution based on its own environment. Each finch perfectly suited its particular island for survival. The birds with genetics less ideal for their location were not able to survive as well as those whose genetics were ideal. This allowed greater opportunity for the properly suited finches to breed and pass their traits down, thus growing the number of birds properly suited for their environment(s) and dispersing the trait(s) throughout the species. It is through this gradual thinning-out process that natural selection provided the environmentally best-suited traits to contribute to the evolution of each type of finch. Darwin's theory applies to all species, and it brings genuine meaning to the phrase "survival of the fittest."

Genetic drift is another phenomenon that affects evolution, in which species lose a certain amount of genetic variation due to specific causations. This loss of variation shifts the direction a species will advance. The most notable form of genetic drift is the bottleneck effect. The bottleneck effect occurs when the population of a species is severely reduced. The loss in population slims down the amount of possible genetic material for reproduction, robbing the species of potential paths of evolution. The remainder of the species is then limited to the genes of the survivors and the future branches of DNA from those genes. In the 1890s, a species of northern elephant seal experienced the bottleneck effect. They were nearly hunted to extinction, but for roughly twenty seals. Once the seals repopulated, the new generations were seen to have much less genetic variation, for they were made from a very small sample of genetics. These elephant seals and all other species who have experienced this effect will prove to evolve into less variable and less differing offspring. The bottleneck effect affects the evolution of certain populations greatly, for it decides the genetic line down which a species will evolve.

Both Darwin's theory of natural selection and the concept of genetic drift have significant effect on the process of evolution, as each helps to decide the genetics of a species' future.

***How has Natural Selection produced drug-resistant bacteria?**

Jeff Ren

Antibiotics have revolutionized medicine and have saved countless lives. However, this defense against bacteria may be soon obsolete as a new generation of bacteria arises. Drug-resistant bacteria are becoming an ever-increasing biological threat. Antibiotics are becoming ineffective as more bacteria gain resistance to modern drugs. In fact, it is estimated that superbugs kill around 700,000 people a year. However, these newer, stronger bacteria are not the product of thousands of years of evolution. In fact, these bacteria first arrived within decades, and have evolved over mere years. How do these bacteria evolve so quickly? The key is with natural selection. Natural selection allows bacteria to gain resistance at such fast rates. These characteristics give bacteria the ability to gain resistance to drugs faster than we can create them.

Superbugs have been evolving since the creation of the very first antibiotics. In the 20th century, the discovery of penicillin and antibacterial sulfonamides gave rise to the modern antibacterial revolution. By 1943, just 15 years after its discovery, penicillin began its mass production. It quickly became one of the most widely used drugs and was considered by many as a miracle. Penicillin was able to treat many previously untreatable diseases for the first time, such as Cholera and Syphilis. However, superbugs soon turned the tide. By 1955, numerous strains of bacteria had already become resistant to the antibiotic. Efforts were made to slow bacterial resistance, such as limiting access to it by making it prescription only. However, these efforts were futile. In 1960, a new antibiotic, methicillin, was created in order to attempt to combat penicillin-resistant bacteria. Nonetheless, within a year, new strains of bacteria gained resistance to methicillin. This new strain, called MRSA, is resistant to most antibiotics, and is common within hospitals. This methicillin resistant staphylococcus aureus (MRSA) kept growing stronger and began infecting healthy people in the 1990s. This made many people aware of this new threat. By 2005, more than 60% of *Staphylococcus* cases were resistant to methicillin, and over 100,000 people were affected by this new disease, 20,000 of which died.

Although superbugs may occur as a natural phenomenon, there are several artificial causes to the creation of superbugs. Superbugs are able to gain resistant qualities through several ways. Natural selection is the process in which beneficial hereditary traits are passed down over time. Organisms often produce more offspring than will survive. The offspring with beneficial traits are more likely to survive and reproduce than offspring without these traits. Since hereditary traits are passed down by the parents, this means more offspring will have the beneficial traits. This process is no different with bacteria, except for the fact that they are able to accelerate the process of natural selection. This is due to their quick rate of reproduction. Each bacterium can split into two every twenty minutes. As a result, more offspring are produced, meaning traits can be passed down quicker. Natural selection explains how these traits are passed down so quickly, but how do these resistive mutations get there in the first place?

The process of mutation still occurs randomly in bacteria. Bacteria, however, are actually able to share genetic material in several ways. This means that resistant qualities can spread not only quickly within a single species, but across many species as well. This is due to horizontal gene transfer. Horizontal gene transfer is the process in which bacteria can share genetic material with other bacteria that are not their offspring. Gene transfer occurs in three different ways. The first method of horizontal gene transfer is transformation. During transformation, fragments of DNA are released from dead bacteria and enter the recipient bacteria. This new DNA fragment replaces an existing fragment of DNA inside the recipient bacteria. The second way gene transfer occurs is through transduction. During transduction, DNA fragments travel from bacterium to bacterium by a bacteriophage. Bacteriophage infect various bacteria and reproduce inside of them. These bacteriophages sometimes form around pieces of bacteria DNA. Whenever these bacteriophages infect another virus, it deposits some of this new DNA in the recipient, where it can be exchanged for a piece of the recipients existing DNA. The third and final way horizontal gene transfer occurs is through conjugation. Conjugation is when DNA passes between bacteria through cell-to-cell contact. During conjugation, pieces of DNA called transposons are able to attach to conjugative plasmids, which are able to transport themselves from bacteria to bacteria. Many of these plasmids can not only transfer DNA to the same species, but also unrelated species as well.

Superbugs are constantly evolving. However, new medicines and methods can be developed in order to combat drug-resistance and its root causes. Although superbugs may occur naturally, there are artificial causes contributing to the increasing number of superbugs today. The main factor contributing, however, is the overuse of antibiotics. Sir Alexander Fleming even warned against drug overuse in 1945, however, many drugs, including penicillin, was overprescribed. The overuse of drugs is the reason why natural selection occurs so quickly in bacteria. These antibiotics kill all the non-resistant bacteria, also killing of competition for bacteria with resistance. After all of the non-resistant bacteria are gone, resistant bacteria reproduce, resulting in a new, primarily resistant strain of bacteria. Overuse of drugs creates resistance, but how can we combat drug-resistance? One main method is to stop reliance on antibiotics. Stopping reliance on antibiotics will decrease the speed at which non-resistant bacteria are killed, and the speed resistant bacteria reproduce. Stopping reliance on antibiotics can be something as simple as washing one's hands to stop the spread of germs. Another method to combat drug-resistance is antisense therapy. Antisense therapy targets DNA in bacteria and prevents gene transcription. This would inhibit the expression of resistant genes in bacteria, allowing the use of existing antibiotics.

Drug resistance in bacteria is an issue that is beginning to become more prevalent in our modern world. The World Health Organization warns that superbugs are becoming one of the most serious threats to global health and security. These bacteria have been on the rise for decades and are evolving at unimaginable speeds. Even though modern medicines have made so many advances, superbugs always seem to be a step ahead. However, new strategies are being put into place to combat superbugs. The phenomenon that are superbugs are truly a wonder of nature.

How did eyes evolve?

Henry Kaplan

The human eye is one of the most complex organs in the entire body. With over 70% of the sensory nerves in your entire body, the eye is the most sensitive organ, excluding the brain. With all these nerves and millions of other tiny parts that make up something that is smaller than a golf ball, one finds themselves wondering how this amazing piece of work came to be. It is a long journey, dating back to over almost 600 million years, during the Precambrian period. Back then, what was known as an "eye" only managed to keep you from running into rocks or other dangerous creatures. Today, the average human can see a single candle flame from over a mile away. What exactly is the cause for this, and how did the modern "eye" evolve from a single patch of nerve receptors?

The eye is the most intricate structure in the body, with millions of tiny photoreceptors that transmit images of everything that is seen. Your eye consists of a few important parts, including the cornea, the lens, the pupil, the retina, and two types of photoreceptors in the retina called rods and cones. Light travels in through the cornea through the pupil, the dark opening in the colored iris, and hits the lens, which then projects the light onto the back of the eye where the retina is located. The pigmented layer of the retina captures light so it doesn't bounce around, and the neural layer after it contains both neural cells and photoreceptors. Photoreceptors, rods and cones, are responsible for the conversion of light energy into images that travel to your brain. There are nearly 120 million rods and about 6 to 7 million cones in each eye. Rods and cones are connected to a bipolar cell through a synapse, and the bipolar neuron connects to a ganglion neuron, which provides a pathway for light. Rods are less developed than cones, and therefore only receive and transmit shades of black and white, and no color. Cones are the ones responsible for sensing color. There are three types of cones: red, green, and blue.

During the Precambrian period, it is suspected that clusters of photoreceptors called “eyespots” were used by animals to detect light. These eyespots were in no way able to be used for vision, as they were incapable to do so. The basic function of these photoreceptor clusters was to detect changes in the amount of light in front of an organism, so as not to run into anything and have a basic awareness of the surrounding area. Eyespots were common to most unicellular organisms, as most multicellular organisms began to develop lenses for a more sophisticated sense of vision. For example, planarians (a species of tapeworm) possess eyespots, which are in cup-like depressions to where it would look like eyes. But soon, larger organisms would need a more sufficient sense of vision to survive.

The first fossil records of eyes date back to the Cambrian period during the “Cambrian explosion.” The Cambrian explosion caused a massive increase in the evolutionary rate in organisms. This evolutionary boost caused an organism to evolve an eyespot into a vertebrate eye (a developed lens but with a somewhat basic sense of vision) in around 360,000 years. Compared to most limbs that evolved in species over the course of millions of years, this is extremely fast for such a finely detailed organ. During those 360,000 years, eyespots transformed into deeper, more eye-like shape with a lens and a larger number of photoreceptors located in a retina. This allowed creatures to have a refined sense of vision and be able to make out distinct objects.

The creation of a lens in the eyes is an evolutionary result in lobopods (a worm-like creature) allowing them to see into darker areas of water. The lens increased the intensity of the light taken in by through the eye, which meant that the lobopods would be able to see more effectively in dark areas. This was the creation of what is known to be as a typical "vertebrate eye." It contained a developed lens that could focus images more clearly and an empty, bowl-like shape instead of a shallow depression, thus creating greater surface area for light to hit. But a new problem arose. The front of the eye needs to be covered in a transparent layer of cells, and this is biologically difficult to do. That is until organisms used the deposition of transparent minerals for both nutrient supply and waste removal. This became what is essentially "blinking."

Finally, an eye complete with a retina and a developed lens had evolved. The only thing left to do now was add the finishing touches. One of these was the development of five opsin classes, which allowed for colors to be differentiated. In land animals, this did not vary by much, but in the ocean, many different species of fish live where there is both a lot of light and no light. As you go deeper into the ocean, longer wavelengths of light like red and yellow get absorbed quicker than the short purples and blues. In order to see prey and predators in deep waters, fish evolved photoreceptors to distinguish colors better with limited amounts of light. When fish moved onto dry land and evolved into land creatures, the path to the eye we know today continued.

In what ways has Darwin's Theory of evolution been proven to be true?

Rohan Reagan

The Theory of Evolution first came around in 1859 when Charles Darwin published his book "On the Origin of Our Species". This theory of evolution was based around natural selection being the reason for an organism's physical and behavioral change over time.

Fossils are one of the most useful ways to prove Darwin's Theory. A thousand years ago, species looked extremely different than ones today which is apparent through fossils. The various prehistoric species such as a *Rodhocetus*, which whales are descendants of, exhibit this. Scientists discovered this species and saw that it had a pelvis and hind legs. The only difference is that the pelvis was not fused to the backbone, so the animal swam. This evidence shows the whales used to be land mammals but evolved into the sea animal that we know today. Whales also contain a bone structure in the forelimb which is very similar to humans, birds, and dogs. This is called a homologous structure because it is a bone structure shared due to a mutual ancestor. This evolution of the bone structure of the whale is noticeable, but there is much more going on that we cannot see.

Biomolecules have become the most reliable way to support evolutionary theory. The basic idea behind this thought is that all living things use DNA and RNA which suggests that all living things are related. This relationship can be shown through the genes, or protein encoded by amino acids, of humans compared to others. An animal very similar to us, the chimpanzee, has genes that match 98.6% of the ones of humans. A common misconception is that humans evolved from chimpanzees or apes but this, however, this is not true. Instead, both chimpanzees and humans, years ago, came from a common ancestor. Other examples are mice which match around 85% and fruit flies which match at approximately 50% to humans. This greater difference in genes shows how they are not as closely related to humans as other animals.
Microevolution can be shown easily through direct observation. In 1959, a village in India experienced a species growing resistance to the chemical DDT, used to kill mosquitos. At first, the DDT killed around 95% of the mosquitos in the area. When they did this a year later, it only killed 45%. This shows that the surviving mosquitos, the first year that the DDT was used, reproduced passing on the immunity to the chemical. The mosquito's large population size and short lifespan helped contribute to the rapid resistance to the DDT. The mosquitos having a large population allowed them to have a greater chance of having random mutations occur. These two reasons are also why bacteria and viruses can rapidly evolve becoming resistant to drugs.

Evolution is proven in numerous ways in a multitude of fields. Fossils belonging to animals that lived during the prehistoric age show that whales used to walk on land. Biomolecules show how closely related different animals are. Microevolution is shown through the short living species, such as mosquitoes, which can quickly evolve to fit their situation.

***How were the first few species formed?**

Lara Seledotis

The beginning of species formation is a widely debated topic because no one was there to witness how it began. A species is defined as a group of related organisms sharing common characteristics. There are many different theories on how life split into various species. Moreover, species can interbreed and produce offspring.

The first theory that will be discussed is the RNA-World Hypothesis, which involves the replication and mutation of strands of RNA. Millions of years ago, the Earth's environment was able to produce infinite numbers of RNA. RNA is a strand of nucleotides. Each nucleotide has other specific nucleotides it attaches to. Once multiple singular nucleotides pair with a strand of RNA a chemical reaction occurs and forms 2 attached molecules of RNA. This is called base pairing. Once RNA is paired a complementary strand is formed. This strand is an indirect clone of the other strand. If the water RNA is in is heated the base pairs lose their grip and they disconnect into 2 complementary strands of RNA. This process happens over and over again, but occasionally mutations occurred, causing RNA to begin forming new 'clones'. Portions of some strands of RNA attached to itself, forming Ribozymes. These Ribozymes had unique shapes which determined their function. They could either break apart or combine other molecules. Eventually after plenty of mutations a Ribozyme was able to combine random molecules to form nucleotides. This allowed for RNA to have a constant supply of nucleotides and eventually connect to living things.

The second theory to be discussed was based upon the Miller-Urey experiment results. The Miller-Urey experiment demonstrates what the early stages of Earth's atmosphere could have been like. Using this they were able to see how the molecules in the ocean had simple reactions to later form more complex compounds, and eventually, life. During this experiment they made a device to demonstrate what earth's early atmosphere could have looked like. This contained ammonia, hydrogen, and methane. They concluded that in the water over a long period of time the heat and gases on early earth allowed for simple reactions to take place. Eventually, they were able to form more complex reactions and compounds.

Living organisms can become more structurally advanced due to the hypothesis of the bacterial origin of the mitochondria. Mitochondria are located in cells. A Mitochondrion contains DNA, which is similar to bacterial DNA. Also, a mitochondrion has ribozymes and transfer RNA molecule similar to bacteria. But how did mitochondria originate? It is believed that mitochondria back in early earth came from specialized bacteria that survived endocytosis and got incorporated into the cytoplasm. Since the bacteria could conduct cellular respiration in the host cell, relying on the glycolysis and fermentation of the cell, it had an evolutionary advantage. Likewise, bacteria that could conduct photosynthesis also had an advantage, thus the population of these cells would increase greatly. Since these cells were not believed to contain enough DNA to work as specialized mitochondrial proteins, they would have to undergo millions of years of evolution, believed to have eventually formed mitochondria.

Scientists don't know everything about how life started but they have come up with theories and experiments to test them. Even though these theories cannot be entirely proven there still is evidence and experiments to help support these theories. Likewise, just because there is some evidence and experiments backing up these theories does not mean they are true. As a re-cap, the RNA hypothesis involves the cloning and multiplying of simple RNA strands to possibly form more complex self-sustaining Ribozymes with specific functions. The Miller-Urey experiment showed us what earth's early atmosphere could have looked like, and how maybe it could form simple reactions to form simple molecules thus leading to more complicated compounds. Our last theory is how bacteria started life not by itself, but by surviving in a host cell long enough to evolve and increasing its population so over millions of years, it could hypothetically form mitochondria.

Hormones

What are the pros and cons of new reproductive technologies?
Jason Liu

Your child has been born and you want nothing more than to hold them in your arms and watch as they fall asleep. Having a child is the most fulfilling feeling as a parent, but what if you can't have that feeling? What if a woman cannot become pregnant. This is where modern reproductive technologies come in. New ways are being developed to give parents who are unable to get pregnant a chance to witness the birth of their own child. Although that sounds like the perfect solution, there are still some issues that come with it. These are the pros and cons of modern reproductive technologies.

The main benefit from reproductive technologies is that it grants people the ability to have a child. As men and women get older, infertility becomes a greater issue.

In Vitro Fertilization (IVF) or Clomids are two forms of reproductive technologies. IVF is when the egg and sperm meet outside the body and fertilization occurs in a controlled environment. Egg and sperm meet in a fertile environment where the embryo can start developing before being implanted inside the mother. Clomids is the term used for medicines that treat infertility, such as Viagra. They work by tricking the pituitary gland into thinking that the male or female is low on testosterone or estrogen, respectively, and forces the body to produce more of these hormones. Both of these options have very high success rates and are able to give a couple a chance to have their own kids.

Although reproductive technologies appear to be a miracle for some people, something that might deter people are the monetary costs, as well as the costs on one's own body. In IVF procedures, the expenses are high, but there is also the mental stress that goes with it, an example being the worry about failure.

In Clomids, the cost is something like that of melatonin, but it can cause innumerable pains. When one is infertile, things such as the pituitary glands are deficient in hormones and in some cases they are blocked, so when clomids come in to stimulate hormone release, it reverts your body back into the time similar to puberty: full of mood swings and headaches in men, and ovarian cysts and hot flashes in women. Furthermore, these may be accompanied by urinary and muscle pains.

These technologies may also be a good solution towards those with genetic illnesses such hemophilia or genetic abnormalities because the developing embryos are constantly monitored. In the event of an unhealthy embryo, it can be terminated. IVF allows doctors to fertilize many eggs and take the best sample out of the batch. IVF utilizes several sperm and egg cells in order to ensure a child with the fewest issues.

Genetic issues are a concern because babies can be born underweight. Also, because some of these diseases are latent, other genetic diseases such as cerebral palsy may arise. Moreover, sometimes there are variables that cannot be controlled in processes like these. Babies born from IVFs are also called test tube babies because they originate in an unnatural environment. Research is still being done on whether or not there are some issues in the child's later life due to IVFs.

In conclusion, all forms of reproductive technology have their pros and cons. Nothing is perfect, and neither are these. There is a saying that "Nothing great was ever accomplished without making sacrifices," and that is true with this. Reproductive technologies are getting better and better and it will make a positive impact on the world.

What are the positive and negative effects of anabolic steroids in athletes?

Travis Ilitch

While watching sports on television, have you ever wondered how some athletes' muscles seem so impossibly developed?

Naturally, they are impossible to get. Some athletes take a drug called anabolic steroids to help increase the rate of muscle gain in their body to achieve a competitive edge over other athletes. Most do not realize the health risks that come with anabolic steroids or even what they really are.

Anabolic steroids are much more than a clear liquid in a syringe that makes you stronger. Anabolic steroids are lipids but more specifically hormones. A couple other well-known steroid hormones are estrogen and testosterone; the functional hormone in anabolic steroids. One's body has a certain level of testosterone and as one injects anabolic steroids into their bloodstream the artificial testosterone is intended to increase the rate of muscle growth.

There are short-term negative results to abusing anabolic steroids. "Anabolic" means building up, which is precisely what anabolic steroids do. Due to the increased amount of artificial testosterone, when an athlete works out and tears his tissues the artificial testosterone increases the rate of tissues being being rebuilt. Results come easier and much faster to users of anabolic steroids and this gives many athletes edges over others. No matter one's sex, anabolic steroids will create results.

Over longer periods of time the negative effects of these steroids increase. Inside the body one will begin to experience a large diversity of issues. Some examples in the blood include development of cholesterol patterns associated with heart disease, obstructed blood vessels or stroke. In the heart one can experience elevated cholesterol and increased blood pressure. Some may experience impaired liver function, peliosis hepatitis (blood-filled cysts that can lead to liver failure), and tumors on the liver. Along with these negative outcomes, younger steroid abusers can experience stunted growth and an increased rate muscle strains and ruptures. Skin side-effects could occur too, such as increased acne, male pattern baldness, edema (swelling), and striae (stretch marks). Finally, taking anabolic steroids can lead to hepatitis B, hepatitis C, and HIV (if needles are shared). No one experiences the same side effects, but everyone experiences something negative and possibly life threatening.

Along with many physical problems, steroid abusers will begin to experience many psychological issues and gender specific issues. One who abuses anabolic steroids can experience mood swings, aggression, violence, depression, psychotic episodes, and addiction. Most of these psychological negatives can be addressed as '"Roid rage". Females have many gender specific side effects mostly due to testosterone being a predominantly male hormone. These include, increased risk of cervical and endometrial cancer, increased risk of osteoporosis, temporary infertility or sterility, altered sex drive, changes in fat distribution, birth defects in future children, increased facial hair and body hair, deepening of the voice, shrinking of the breasts or uterus, clitoral enlargement, and finally menstrual irregularity. Many effects can occur to males as well, such as temporary infertility or sterility, altered sex drive, prostate enlargement, painful erections, shrinkage of testicles, reduced levels of testosterone, abnormal sperm production, and increased levels of estrogen. With so many physical, gender specific and psychological risks, it makes little to no sense why one would ever abuse anabolic steroids.

Overall, steroids are not worth the risk that comes with them. Athletes don't understand the seemingly countless negatives that come with a slight edge in whatever activity they are participating in. Many athletes are aware of the risks associated with anabolic steroids, but they still abuse them due to addiction. One should never experiment with anabolic steroids because the small bonus it gives in sports is not worth the possibly fatal risks.

Anabolic steroids are a drug that allow you to get stronger and faster, but at a cost that is not worth it ever. The artificial testosterone builds up in the tissues and does cause an increased rate of tissue growth, but the repetitive theme of this written piece is using anabolic steroids is not worth it at all and one should never try or abuse them.

What is the pancreas? What are the hormones of the pancreas, how are they secreted and what are their functions?

Akash Raman

Have you ever wondered why you feel shaky, irritable, or tired after you skip a meal? This is because the sugar level in your blood is low. The pancreas is an organ you've probably never really thought about. However, it is vital for survival as it monitors and regulates the levels of sugar in the blood. When we eat, our body breaks down certain foods into sugar molecules like glucose, which is the main source of energy for our body. It is important to remember that glucose cannot enter the cells of tissues and organs without assistance from hormones secreted by the pancreas. This vital organ is not very big. It is in the abdomen, situated in the upper left corner. It's size is about one inch in width and seven inches in length. The head of the pancreas is situated in the curve of the duodenum and its tail is between the stomach in front and the spine or vertebra behind.

The pancreas secretes hormones, which are chemical messengers. It is two main types of glands - endocrine and exocrine. The exocrine glands secrete substances through tubes or ducts. Endocrine glands do not release their secretions, hormones, into ducts, but directly into the interstitial fluid, which are then picked up by the blood. A German pathologist discovered specialized cells in the pancreas that functioned as endocrine glands called alpha cells and beta cells. Alpha cells secrete glucagon and beta cells secrete insulin. Glucagon and insulin work opposite to each other in order to control the balance of glucose in the bloodstream. When the blood sugar level is low, the pancreas secretes glucagon. In the liver, this glucagon helps the liver to use the stored glycogen in its cells and make glucose. Glucose is then released into the bloodstream. As a result, this hormone thus helps to raise blood sugar levels. However, as soon as the pancreas senses that the glucose levels in the blood are too high, it produces insulin. Insulin makes glucose enter the liver cells away from the blood. The arrival of glucose activates glucose transporters in the cell membranes. These transporters signal glucose to enter into the cell. Insulin lowers blood-glucose levels by causing glucose to be stored as glycogen in muscle and liver cells. This glycogen can be used for energy when needed.

The pancreas secretes hormones responsible for regulating blood sugar levels. These hormones ensure that sugar levels do not drop between meals. The pancreas secretes another hormone, somatostatin, to help maintain a balance of sugar in the body. Somatostatin blocks the secretion of both Insulin and Glucagon. Pancreatic Polypeptide is another hormone that self-regulates the pancreatic secretory function.

In short, the pancreas is a very important organ, vital to life that cannot be forgotten given its small size. Each hormone secreted by the pancreas has specific and important functions that help maintain homeostasis in our body. Size can never determine importance.

What is the role of the parathyroid glands, and what conditions can result if they function improperly?
Zach Weinstein

In the neck there are four disc-shaped endocrine glands. These glands are responsible for calcium regulation in the body. They can also develop conditions that are very dangerous. They are called the parathyroid glands. The parathyroid glands are endocrine glands, meaning they are ductless, secreting Parathyroid Hormone straight out of the Parathyroid itself. This Parathyroid Hormone, referred to as PTH, travels through to body and eventually reaches the bones which it breaks down to release calcium. This journey in order to produce calcium is very intricate and will be explored in further detail.

The four disc-shaped, rice sized parathyroid glands are located in the neck behind the thyroid. These parathyroid glands are endocrine glands, with the very important role of calcium production. They secrete PTH directly into the blood. The PTH is carried through the bloodstream until it reaches one of two places, having three possible effects. As discussed before it is possible to the PTH to break down bone matter to release calcium. It does this by binding to osteoblasts, the cell responsible for creating bone matter. When bound to by PTH the osteoblasts produce extra RANKL. RANKL is a protein responsible for bone generation. Osteoblasts are now inhibited from producing osteoprotegerin (OPG). The remaining OPG compete to bind to RANKL. The RANKL will bond to RANK, also responsible for bone generation. This bonding of RANKL to RANK stimulates the production of osteoclasts. These osteoclasts will then break down bone material and release calcium into the blood. This one of the ways PTH is used, another is when PTH travels to the kidney, where calcium is absorbed, and increase the rate at which it is absorbed. The last role of PTH is to stimulate the production of active Vitamin D which will increase the rate at which calcium is absorbed in the intestine. In summary the parathyroid produces PTH which in many ways is essential to production of calcium in the body.

The parathyroid glands can develop a few conditions that negatively affect the body including hyperparathyroidism, hypoparathyroidism, and parathyroid cancer. Hyperparathyroidism is a condition where the body produces too much PTH. It is most common in women around the age of 65. This leads to a few negative effects. One is that the body will take too much calcium out of the bones. This is dangerous because it can lead to osteopenia, thinning of the bones. It can even lead to osteoporosis if left untreated for a long time. Since the body is absorbing so much extra calcium it is likely that the body will develop hypercalcemia. This leads to high blood pressure, kidney stones, fatigue, and memory loss. Luckily, this can be treated with a fairly simple surgery, called a parathyroidectomy. A parathyroidectomy involves removing one or more of the parathyroid glands to normalize PTH and calcium levels. The recovery for this surgery is very high and little risk is involved. On the opposite side is hypoparathyroidism, where the body doesn't produce enough PTH, causing hypocalcemia. In extreme cases muscle cramping, weakness, confusion, exhaustion, and tingling may be present. This can be helped with a simple calcium supplement. These issues are rare and have fairly easy fixes, but one issue that is extremely rare but also very serious is parathyroid cancer. Parathyroid cancer will display the same symptoms as hyperparathyroidism, but to a much more extreme level. Usually the only fix is to have a removal of both the parathyroid and thyroid. This will require many supplements to be taken for the rest of the patient's life. These issues are serious, but fortunately science is evolving and we will know more possible cures as time passes.

The parathyroid glands have an essential role in the body and can even cause the body some issues. These small glands are essential to calcium maintenance. They use PTH to affect bones, kidney, and small intestines. These little glands can also cause your body some issues, luckily fixes are known for all common issues.

What is the hormonal and neurological basis of emotions?
Anya Tarasovets

The hormonal and neurological basis of emotion is very complex and requires multiple parts of the human brain to work together. When all of these layers collaborate, the human body experiences a variety of emotions. Emotions that a person might feel are based upon many factors, including their bodies and mind's reaction to a given scenario and the way their mind processes these scenarios based on prior knowledge and memories. The mind's reaction to these events causes the release of hormones which help drive emotions and change the body physically.

To experience emotions to their full potential the human mind must first process current events and decide how to respond. When a person is born they have no memories. Although this is true, as the baby grows their brain develops and begins to associate scenarios with emotions and create memories. These memories are key in helping the brain understand the emotional state that a human might be in. Once the brain recognizes the situation and associates it with a memory it sends signals to the cerebellum. In the cerebellum, different hormones are released that cause physical changes in the body. Chemicals like dopamine, serotonin, oxytocin, and endorphins cause feelings of happiness and make the human body full of energy and excitement. These different chemicals are the reason that emotions are different for many people. The feeling of being in love is mainly produced by norepinephrine which speeds up heart rate, dopamine that makes a person feel joy, oxytocin that releases the desire for closeness, and phenylethylamine that generates the need for intimacy. Humans summarize all these complex chemicals into one simple word: love.

The human body responds to current events by understanding them and sending signals to the amygdala and hippocampus. When going through an event, the amygdala must recognize key ideas in this event and send signals to the hippocampus. The amygdala is responsible for the emotional responses of the brain. It picks up major ideas and relates them back to a memory. The memories are stored in the hippocampus part of the brain. The hippocampus relates all memories back to a key emotion. For example, the aftermath of a tragic event can cause one to feel great sadness and grief. These labeled feelings help describe what that person might be going through. Death is universally known to cause mournful emotions, because the hippocampus stores all of the memories of death that one person had and compares them with their response. The death of a family member usually brings tears which is the body's reaction to that scenario. The hippocampus stores this reaction, so that next time death comes up the person's body will recognize this similar scenario causing the hippocampus to trigger melancholy emotions.

Every person is unique in the fact that any daily moment might be associated with a memory from their past. The amygdala part of the brain takes this memory, which is stored in the hippocampus, and sends it to the cerebellum. Finally, the cerebellum sends out different chemicals to the cortex and the cortex projects them onto the human body. In conclusion, all of these parts of the brain are working hard in order to make a person feel anything, from anger to anxiety.

*What hormones influence human reproduction?

Joshua Zexter

Humans reproduce to keep their species alive. In the past century, physiologists have finally discovered why the phenomenon has continued happening for so long. Hormones, which are chemical messengers that cause specific changes in the body, were discovered. Hormones such as testosterone, estrogen and progesterone, oxytocin and vasopressin play roles in reproduction.

Testosterone is the primary male hormone and is an anabolic steroid. In male humans, testosterone plays a key role in the development of male reproductive organs such as testes and prostate, as well as promoting secondary sexual characteristics such as increased muscle and bone mass, and the growth of body hair. In addition, testosterone is involved in health and well-being and the prevention of osteoporosis. Insufficient levels of testosterone in men may lead to abnormalities including frailty and bone loss.

Estrogen is the primary female hormone. It is responsible for the development and regulation of the female reproductive system and secondary sex characteristics. There are three major estrogens in females that have estrogenic hormonal activity: estrane, estradiol, and estriol. The estrane steroid estradiol is the most potent and prevalent of these.

Progesterone (P4) is an endogenous steroid involved in the menstrual cycle, pregnancy, and embryogenesis. It belongs to a group of steroid hormones called the progestogens and is the major progestogen in the body. Progesterone has a variety of important functions. It is also a crucial metabolic intermediate in the production of other endogenous steroids, including the sex hormones and the corticosteroids, and plays an important role in brain function as a neurosteroid.

Oxytocin (Oxt) is a peptide hormone and neuropeptide. Oxytocin is normally produced by the hypothalamus and released by the posterior pituitary. It plays a role in social bonding, sexual reproduction, childbirth, and the period after childbirth Oxytocin is released into the bloodstream as a hormone in response to stretching of the cervix and uterus during labor and with stimulation from breastfeeding. This helps with birth, bonding with the baby, and milk production. Oxytocin was discovered by Henry Dale in 1906. Its molecular structure was determined in 1952. Oxytocin is also used as a medication to facilitate childbirth.

Hormones are the main force that drives human reproduction. Testosterone plays the key role for males while estrogen and progesterone play the key roles for women in reproduction. While Oxytocin is present in both males and females, it is predominant in females. Without these main hormones, human reproduction would either be impossible or just wouldn't happen.

Circulatory system

**How does the human heart function?*
Beckett W

The human heart is a central organ of the human body. Though it is only the size of a clenched fist, the heart has many roles that are essential to life. In order to understand the heart's job, it is necessary to know its anatomy and function. The human heart is made of 4 chambers which contract and push blood throughout the body using vessels. The heart beats because of an electrical impulse that is created by the Sinoatrial node.

The human heart is muscle located in the chest, behind the sternum and above the diaphragm. This mass contains four large chambers, which have extremely important jobs. There are two upper chambers, which are smaller, called atria which hold the blood before it is pushed through a valve. The Mitral valve is located on the left side of the heart and the Tricuspid is located on the right. When the valves open, the blood moves from the atrium down into the larger chambers called the ventricles. The ventricles then contract and push blood through vessels that branch off from the heart.

There are two different types of vessels that are connected to the human heart, arteries and veins. Arteries transport blood away from the ventricles after ventricular contraction. The aorta and pulmonary arteries are the two major arteries. The aorta takes oxygenated blood from the left ventricle and transports it to most of the body. The pulmonary artery transports deoxygenated blood to the lungs. Though the arteries are able to take blood away from the heart, blood still needs to return through another type of vessel.

Veins are the other type of vessels connected to the heart. The veins transport blood back to the heart from different parts of the body. After the blood makes its way back to the heart from the most of the body, the blood, usually deoxygenated, flows into the right atrium through the superior and inferior vena cava. Blood from the lungs returns the heart by the pulmonary veins. This blood is usually oxygenated and goes to the left atrium. All this blood transportation needs to start with a small electrical impulse.

An electrical impulse is needed to cause the heart to contract. The sinoatrial (SA) Node generates this impulse and is located in the upper part of the right atrium. This electrical impulse makes its way down to the atrioventricular (AV) node located in the lower part of the right atrium. After the AV node, the impulse moves into the bundle of his, a collection of heart muscles that conduct electricity, located in the upper part of the ventricles. The bundle of His then branches off into the left and right bundle branches which stretch throughout the muscular wall between the two ventricles. From these bundle branches, the charge moves into the Purkinje fibers that penetrate and surround the ventricles and cause the heart to contract. This impulse has found its way from the SA node to the Purkinje fibers and causes the left and right ventricles to contract and push the blood through the pulmonary and aortic valves and into the arteries.

The heart is the one of the most important organs in the body. But the heart would not be able to function without its essential parts. These include the four chambers, two atria and two ventricles, and the vessels that branch off the heart, veins and arteries. Along with, the electric impulse creator, the SA node, and the impulse transporters, the AV nodes, bundle of His, bundle branches, and Purkinje fibers. When all these parts work together, they give the human heart its function.

What is atherosclerosis?

Helen Qin

Atherosclerosis occurs when plaque, a material consisting of fat and cholesterol, builds up on the walls of the blood vessel. This build-up of plaque causes the lumen of the vessel to narrow and the vessel to harden. Atherosclerosis is known to be a slow and silent killer, since it usually does not show any symptoms until one has reached old age. It is also responsible for heart attacks, strokes, and many other cardiovascular diseases.

Atherosclerosis begins when the endothelium is damaged. The endothelium is a thin single cell layer on the arteries and blood vessels that keeps it smooth. Endothelial damage has many causes including high blood pressure, smoking, and high cholesterol. Plaque forms due to this endothelium damage. This plaque formation occurs when LDL, also known as "bad cholesterol", passes through the damaged endothelium and enters the wall of the artery. Later on, when the white blood cells start to eat away at the LDL, plaque is formed.

There are many health problems linked with atherosclerosis. In fact, most cardiovascular diseases have a connection to atherosclerosis. For example, due to the narrowing of the blood vessel caused by plaque build-up, often a piece of plaque may crumble off the blood vessel walls and travel down the blood stream until it reaches the smaller blood vessels by the brain where it will clog it. This is also known as a stroke. If enough plaque builds up, it can cause a blockage that will affect the entire body negatively. Atherosclerosis can also cause heart attacks, peripheral vascular, and many other cardiovascular diseases. When the blood vessel lumen becomes too narrow, one may experience pain. Sometimes, blockages can even rupture and cause blood clots at the site of the rupture.

However, many things can prevent atherosclerosis including a low intake of fat and LDLs. Not smoking and exercising regularly can also help prevent atherosclerosis. Maintaining a normal level blood pressure and cholesterol level also decreases your chance of atherosclerosis. However, diabetes, stress, and abdominal obesity can cause atherosclerosis. A diet lacking fruits and vegetables coupled with excessive alcohol intake can further increase one's chance of atherosclerosis. All the factors mentioned can lead to many cardiovascular problems. However, all these factors can be prevented. By preventing these factors, you reduce your risk of atherosclerosis and many cardiovascular problems.

Atherosclerosis is when one has a buildup of plaque that narrows and hardens their blood vessels. It is caused by many factors including a diet high in fat and cholesterol. Most Cardiovascular diseases have a connection to atherosclerosis, and some cardiovascular problems are caused by atherosclerosis. As many of the main causes of atherosclerosis and cardiovascular disease can be prevented, there are many steps one can take to greatly reduce the risk of cardiovascular problems.

What are the causes of heart attacks?

Praneel Pillarisetty

A Myocardial Infarction, more commonly known as a heart attack, is the damage or death of cardiac muscle tissues. Heart attacks occur when blood flow to the heart is blocked by substances such as fat and cholesterol. This forms a plaque which leads to a clot developing in the arteries. Blocked arteries lead to a lack of blood reaching the heart which can damage or destroy some of the cardiac muscle, causing a heart attack. When cardiac muscle tissue is destroyed, blood isn't pumped at a high enough pressure to reach certain parts of your body, such as the brain. If oxygen doesn't reach the brain, it can result in a stroke. This shows that being prone to heart attacks can lead to many more health risks in the future. Heart attacks are mainly caused by actions we make in our life such as our blood pressure, smoking, and use of certain drugs. However, there also are risk factors of suffering a heart attack that are out of our control including family hereditary and disorders.

High blood pressure, high cholesterol, and obesity are a leading cause for a heart attack. If a person has high blood pressure, increased pressure of blood is flowing through the arteries. This increases the amount of fatty materials and cholesterol that can build up in the arteries. Once the arteries become filled with a lot of plaque, oxygen in the blood isn't able to reach the heart muscles, causing a heart attack. To prevent high blood pressure and cholesterol, a person can exercise and eat a healthy diet, preventing buildup of unnecessary material in the arteries.

Smoking releases nicotine, an addictive chemical that can cause a buildup of plaque in arteries. Someone who smokes is twice as likely to have a heart attack in the future compared to a non-smoker. First, the carbon monoxide in tobacco reduces the amount of oxygen in your blood, forcing your heart to pump harder to supply the body with the oxygen the body needs. Also, the nicotine stimulates adrenaline, making your heart beat faster. This increases blood pressure significantly, which is a risk factor that causes heart attacks. If a non-smoker breathes in secondhand smoke, it can also be harmful towards them. Research shows that this is a cause of heart disease in non-smokers. Overall, smoking raises your blood pressure significantly, which leads to heart diseases.

The use of certain drugs can clog up your coronary arteries, reducing blood flow to the rest of the body. Cocaine, an illegal drug in the United States, is used by many people. It is a stimulant that people use to raise their alertness. This increases your blood pressure making your chances of suffering a heart attack higher. Other drugs such as amphetamine and ecstasy also act as stimulants. Amphetamine is used to treat Parkinson's, obesity and Attention Deficit Disorder. It stimulates your nerves and your brain, also raising blood pressure. Overall, certain drugs can cause an increase in blood pressure making a person susceptible to heart attacks.

There are also risk factors of suffering a heart attack that are beyond your control, such as your family heredity and disorders. It is common for people to suffer heart attacks when many of their family members suffer heart attacks. Genetic factors can play a role in blood pressure, heart disease and other related conditions. Also, certain disorders can make a person more prone to heart attacks than others. These disorders can increase blood pressure and increase plaque buildup in your arteries. These are certain risk factors that are out of your control but maintaining a healthy lifestyle will lower your chances of suffering a heart attack.

In conclusion, a heart attack can occur from choices you make and risk factors that are out of your control. Heart attacks occur when plaque builds up inside your arteries and prevents blood flow to the heart, causing the death of muscle. There are many causes that increase the amount of fatty materials and cholesterol in your arteries and maintaining a healthy lifestyle will help you avoid heart attacks.

***What are the psychological effects of heart disease?**

Julia Frykman

Many people focus on the physical effects of heart disease, but what are the psychological effects? Some patients and survivors of cardiac events may never have any psychological effects, whereas many may find themselves with depressive symptoms. Many patients may never get the right treatment or even tell their loved ones, which can be fatal.

Patients who have coronary heart disease or have had a cardiac event, such as a heart attack, often experience some levels of depression and anxiety afterwards. Cardiac episodes don't only take a toll on your body, but they can also cause PTSD to develop. Having a cardiac event can make someone feel weak and not in control of their own body, which is where the depression stems from. The trauma of an event can cause the anxiety and depression to work hand in hand to further harm someone's body. Women who have had a cardiac event are two times more likely than men to develop depression and anxiety. After a cardiac event, patients have a fifty percent chance of developing depression and of that fifty percent, twenty percent of patients will develop major depression, a depression that causes thoughts of suicide, loss of interest, weight gain, and more.

With depression and anxiety comes an increased risk of a relapse of a cardiac event. Complications such as an increased heart rate and increased blood pressure are some of the effects of having anxiety. With these conditions comes a elevated risk of having another cardiac event, which can lead to death. Many patients also experience panic attacks and fear that they are having another heart attack or a cardiac episode. In return, patients become more anxious and more problems can arise.

These episodes can make it hard to get back to your normal life and patients may suffer from insomnia. Having anxiety while also having cardiac disease can cause the recovery time to increase. It makes reconnecting with relatives and sleeping difficult, along with also making the return to work difficult. Additionally, women are two times more likely to develop depression and anxiety.

Patients that do not have a support system or get the correct treatment have shown to be susceptible to more events and death. Patients who suffer from anxiety and depression and do not have a support system often cannot work through the anxiety and depression alone. The side effects of not having friends or family to help is another cardiac event that possesses an increased chance of mortality. The importance of having a support system is overlooked when it could save someone's life. The treatment for depression and anxiety can be as easy as talking with a trusted friend or family member, but in a more severe case, talking to a therapist would be needed.

Depression and anxiety caused by heart disease are pushed aside when they need to be talked about in order for the patient to return to a normal life. The longer these effects go untreated, the higher risk a patient puts themselves at to have another event and increase the risk of fatality. The psychological effects of coronary heart disease cannot be overlooked and need to be treated along with their physical recovery.

What factors may lead to heart disease and what is going on in your body when you have it?

Nishant Shishu

Heart disease, also called cardiovascular disease, is a condition that includes diseased vessels, blood clots, and structural problems in the heart. There are many types of heart diseases, the most common being coronary artery disease. High blood pressure, cardiac arrest, congestive heart failure, arrhythmia, peripheral artery disease, stroke, and congenital heart disease are other common heart diseases. All of these can cause death.

There are many factors that can cause these different diseases and many different things are happening in your body when you contract these diseases.

During the course of coronary heart disease, there is a buildup of fatty substances in the coronary artery, which becomes a plaque. This plaque blocks the coronary artery off and stops blood from moving. Since blood is being blocked, that part of the heart muscle will die. If you have high blood pressure, or hypertension, the force of the blood against the artery walls is very high. If someone has congestive heart failure, their heart does not pump blood as well as it should. When this happens, blood and fluid back up in your body, making you hold on to a lot of fluid causing swelling. When someone has arrhythmia, their heart has irregular beats, sometimes too fast and sometimes too slow. Another form is peripheral artery disease which is a condition where narrowed blood vessels reduce blood flow to the limbs. A stroke occurs when the brain is damaged because of its interruption of blood supply. The brain cells get no oxygen because no blood is flowing there, so the cells begin to die. The final heart disease is congenital heart disease. It is a heart defect that occurs at birth, but it can resurface later in adulthood. Defects can be simple, causing almost no problems, or they can be life-threatening. There are many ways to contract these diseases.

There are many causes of heart disease, so people must be careful about what they are doing. If people in your family have had any of these heart diseases, it is more likely for you to get it. Another main cause is smoking since the blood vessels will narrow. This means little to no blood will reach certain parts of the body and the muscle will die. Obesity is also a risk factor for getting heart disease because it increases strain on the heart. Obesity can cause high blood pressure, high cholesterol, and diabetes. Cholesterol can be affected by age, gender, your diet, and heredity. Diabetes increases the risk of heart disease because it affects cholesterol and triglyceride levels. Another main risk factor is not being physically active. These are the main risk factors for getting heart disease. If you get coronary artery disease, many things can happen.

One of the most common types of heart disease is coronary artery disease, which can cause many other serious heart problems. One thing it can cause is stable angina which is okay at rest, but when the heart rate goes up, not enough blood can get to the muscle because of the plaque. Coronary artery disease can also cause acute coronary syndrome which is broken down into unstable angina and heart attacks. During unstable angina, blood clots form on a plaque. Blood flow can be lost at any time, no matter what you are doing. During a heart attack, either the blood clot blocks the vessel, or part of the plaque will break off and block a smaller blood vessel. Oxygen gets completely cut off and that part of the heart dies. This is how coronary artery disease can cause more harmful things.

In conclusion, there are many factors that contribute to heart disease. Some of these factors are smoking, heredity, obesity, high blood pressure, diabetes, high cholesterol, and many others. The blood vessel being affected has a plaque in it, causing the blood to be cut off. This can be deadly, so make sure you do not do anything that can cause heart disease.

Disease

Who were the main players in the story of antibiotics, and how do antibiotics work?
Nathan Ouyang

Antibiotics, since their inception thousands of years ago when the ancient Chinese discovered that rubbing certain types of mushrooms on them would provide health benefits, have gone from their humble beginnings to a place in society where things like "penicillin" are common household names. Antibiotics, by design, stop bacterial infection, and only bacterial infection, as they were created to stop the rampant bacterial disease epidemics of the early 20th century. These life-saving drugs have been around for thousands of years in some form or another, but their modern forms have only been around for less than a hundred years.

The story of antibiotics cannot be told without describing their nemesis, bacteria and their story, beginning with the discovery of cells by Robert Hooke. Robert Hooke, a British scientist in the seventeenth century, was the first to discover what he called cells. Two hundred years later, Louis Pasteur was the first to prove that bacteria, not the miasma theory of the time, which suggested that a malevolent gas called miasma caused disease, was the primary inducer of disease with his famous swan-neck flask experiment. A bacterium is a simple, single celled organism that can be found everywhere, and most are harmless to humans, but those that are harmful cause disease that can be widely spread and deadly. These bacteria spread from carrier to carrier, leaving a black mark and killing along the way. By the early 20th century, millions were dying each year from the poor living conditions in combination with the many, many deadly diseases that were running around unchecked.

While Penicillin came earlier, it needed the path to be paved by Prontosil, the first commercially available antibiotic, which was pioneered by Gerhard Domagk, a German scientist. Gerhard

Domagk was one of many scientists employed by the German giant IGF (I.G. Farbenindustrie). In the December of 1935, Domagks' daughter tripped down the stair and punctured her hand with a sewing needle and came down with a deadly streptococcal infection. At the time, after the bacteria started multiplying, there was no known counter and, unchecked, his daughter would certainly die. However, at work, he had discovered a drug that would kill the streptococcus bacteria in mice but would fail to work in a chemical bath. Armed with this miracle drug, he violated almost every code, medical or scientific or research imaginable and snuck home with it and injected his daughter. After a few weeks, his daughter miraculously made a full recovery. It didn't take long for IGF and Domagk to extend their patent for the miracle sulfonamide, prontosil, and ship it to market as a drug. It flopped, but it caught on when it saved the life of FDR's son, thus paving the way for Penicillin.

However, the big breakthrough in the young antibiotic field came fortuitously when Alexander Fleming happened upon his bacteria-killing fungus. Alexander Fleming returned from vacation in 1928 to his disorganized laboratory only to find a colony of mold growing on one of his bacterial petri dishes, and near the mold, the entire staphylococcus colony was undergoing lysis, the disintegration of the cell membrane. This incredible fortunate discovery was then carefully studied and eventually, around the 1940's, the public could get their hands on this "miracle drug," as it was hailed by many. Alexander Fleming, along with two other scientists influential to penicillin, would win the Nobel Prize in 1945 for their work.

There are many different types of antibiotics, and can be broken down into many, many different categories, but the method by which antibiotics kill bacteria is divided into two groups: bacteriostatic, and bactericidal. These two types of antibiotics kill in very different ways. bacteriostatic drugs kill by preventing the replication of bacteria, whereas bactericidal kill by straight up killing the bacteria. For example, a bactericidal drug such as penicillin will kill by performing lysis on bacteria membranes, destroying them and letting the bacteria die while a bacteriostatic drug such as prontosil, will disrupt the splitting of the bacteria by preventing the creation of a certain biomolecule and preventing disease that way.

Antibiotics, to this day, are still carrying out their duty to combat bacteria, but as bacteria evolve, so too must antibiotics, or the antibiotics of yesteryear will no longer stop the bacterial infections of tomorrow. Antibiotics have been going through a stagnant era, as no new major categories of antibiotics have been discovered since the 1980's, and while there are experimental drugs in development, it is uncertain whether any of them well ever make it to market. However, bacteria are constantly evolving at a rapid pace, and soon the antibiotics that we have today will no longer prevent bacterial disease. Alexander Fleming noticed that even in a few years that certain strains of staphylococcus bacteria were resisting penicillin. New antibiotics must be developed for the bacterial infections of the future, otherwise the widespread epidemics of the early 20th century will occur again.

***What is pyelonephritis?**

Stephanie Nemeh

Pyelonephritis is a type of kidney infection that can cause severe damage to one or both of the kidneys. It is very common, and if caught early on, it can be treated very easily with medications. In some extreme cases, it can lead to severe kidney damage and possibly result in kidney failure. This infection is caused by different types of bacteria that can spread to the kidney in two different ways.

Pyelonephritis can either be ascending or hematogenous. Most cases are ascending. The bacteria enters the urethra where it spreads into the bladder and eventually travels up to the kidneys through vesicoureteral reflux. Vesicoureteral reflux occurs when the vesicoureteral orifice, a one-way valve that allows urine to flow from the ureter to the bladder, fails due to either a defect or a bladder outlet obstruction. The bladder outlet can become obstructed because of an increase in pressure in the bladder that distorts the valve. This could possibly lead to urinary stasis, which is when the urine stands still in the bladder, making it easier for bacteria to grow. Another less common cause of pyelonephritis is a hematogenous infection. This is when bacteria spreads through the bloodstream. It is caused by septicemia, more commonly known as blood poisoning, or endocarditis, an infection of the inner layer of the heart.

There are a few types of bacteria that can cause this infection. The most common ones for ascending infections are *E.coli*, *Proteus*, and *Enterobacter*. These types of bacteria are all common in the bowel flora. The most common types of bacteria from a hematogenous are *Staphylococcus* and *E.coli.*

The infection can be diagnosed early on from the symptoms. If the infection is ascending, one may experience symptoms such as burning sensations or pain when urinating, bloody or cloudy urine, or a frequent urgency to urinate. Other symptoms include sharp pains in the upper right or left flank, fever, and chills. Symptoms will normally start out as burning when urinating and will progressively get worse as the infection spreads to the kidney.

Pyelonephritis can be treated very easily if it is caught early. It can be treated with a simple medication such as Bactrim. Drinking a sufficient amount of water can also help prevent and treat the infection because it flushes out the kidneys. In addition to that, cranberry juice is also effective because it helps prevent bacteria from sticking to the walls of the bladder. If it is not caught early enough, some complications could occur such as scarring of the kidneys and septicemia. To treat more severe infections, antibiotics through an IV may be necessary. Kidney disease, hypertension, or even kidney failure could occur if the kidneys develop scarring. Septicemia could occur because the infected kidney would not be able to filter the blood correctly, causing it to send blood filled with bacteria back into your bloodstream.

Being female increases the risk of getting pyelonephritis. The urethra of a woman is shorter than the one of a man, making it easier for bacteria to enter the bladder. The urethra is also very close to the vagina and the anus, making it much easier for bacteria to spread. Pregnant women are also at a higher risk of getting a kidney infection that could result in delivering underweight babies.

Pyelonephritis is a very common kidney infection that can be treated easily but needs to be caught in its early stages. There are two types, ascending and hematogenous, and there are a few different types of bacteria that can cause this infection. It can be diagnosed early due to some apparent symptoms, and if caught right away can be treated with a simple medication.

What are ataxias, and what are the different kinds?
Ryan Bhola

Ataxias are often unknown due to their rarity. If anyone were to discover a loved one or themselves having an ataxia, it would be an excellent idea to know more about their problem. Ataxias are a group of diseases, and some do not have a cure. Finding out what these specific kinds of ataxias are can give a new understanding to people interested in ataxias.

An ataxia will always have physical impairments. Some of these burdens may be slow movements, uncontrollable flinching, speech impairments, eye flickers, or shaking. Additionally, difficulty in swallowing, walking, and maintaining balance can also appear. A deterioration in walking may cause an affected person to require a wheelchair. Although these are horrible conditions to have, an ataxia does not cause any pain. Aerobic exercise, physical therapy, and speech therapy can help with dealing with the physical effects of ataxia.

One way an ataxia can occur in the human body is through hereditary means. There may be a defective gene in the body and this can lead to the production of unnatural proteins. These proteins negatively affect nerve cells' function, leading to a deterioration in the cerebellum and the spinal cord. Progression of the ataxia leads to worsening of coordination. However, it is possible that an ataxia can occur in the body without any history in a family with genetic mutations.

An ataxia acquired through hereditary means can also be autosomal dominant. This means the ataxia comes from a dominant gene in one parent, thus the reason why “dominant” is in the name. Autosomal dominant forms include spinocerebellar ataxias and episodic ataxia. Spinocerebellar ataxias affect the cerebellum and sometimes the spinal cord. They are classified by the mutated gene that causes that ataxia. There are more than 40 identified mutated genes, and sadly the number is not stopping. Episodic ataxia regard ataxias that are not progressive, but rather episodic. These include EA1-EA7. EA1 and EA2 seem to be the most common. Additionally, Episodic Ataxia does not lower a person’s lifespan, and can be treated by medications.

In contrast, an ataxia can be autosomal recessive. A recessive gene in both parents leads to an autosomal recessive ataxia. Some autosomal recessive ataxias include Friedreich’s Ataxia, Ataxia-Telangiectasia, Congenital Cerebellar Ataxia, and Wilson’s Disease. Friedreich’s Ataxia is the most common hereditary ataxia and signs and symptoms will appear before the age of 25. Symptoms may include deficient muscle control, diminished swallowing, slurred speech, scoliosis, and foot malformations. An early treatment of heart problems can improve overall life and the rate of survival. Telangiectasia is much less common than Friedreich's, and it is progressive, starting in childhood. On the other hand, Congenital cerebellar ataxias appear at birth and are caused due to damage to the cerebellum. Wilson’s Disease affects people as copper builds up in their brain, liver, and other organs causing neurological issues, such as ataxias.

Perhaps the most important question would be what the causes of ataxias are. As mentioned earlier, due to a recessive gene in both parents or a dominant gene in one parent, there can be an autosomal recessive or dominant ataxia. Ataxias may also be caused by stroke, cerebral palsy, infections, and tumors. The degeneration of the spinal cord and cerebellum can also cause an ataxia.

Ataxias can be complex and unique, with multiple effects and causes. However, they all have one thing in common: physical impairments that become a heavy burden. Additionally, some ataxias may be passed on hereditarily, causing autosomal dominant or recessive ataxias. Multiple and unique causes lead to ataxias, with some causes having the ability to be responsible for more than one disease.

How do bones heal after a fracture?

David Safta

Bones hold us together, but they also break us apart. The structure of the human body is dependent on bones to function. Bones are composed of a protein called collagen and calcium phosphate. Collagen is a soft living tissue that provides soft structure for the bone, and the calcium phosphate provides strong support to the bone. Bones fracture when the amount of pressure on the bone exerts the bones breaking point. Bone fractures can cause internal bleeding, hemorrhaging and death.

The bones in the body are strong and able to support hundreds of pounds, but they are not invincible. Bone fractures are common and effect thousands each year. Fractures occur when the amount of force exerted on a bone is stronger than the bone itself. Fractures can also occur after a bone undergoes repetitive stress which causes the bone to eventually fail and fracture, or it causes bone weakening. Diseases like osteoporosis weaken the bone and can eventually cause a fracture. Bone fractures can cause nerve damage, infection, loss of mobility and internal bleeding.

While bones are extremely strong they are not immune to breaks. Luckily, the human body has a advanced system to repair and heal bones, called primary healing. Primary healing occurs in five phases, the inflammation, soft callus, hard callus, ossification, and remodeling phase. The inflammation phase occurs right after the fracture begins as the bone periosteum (the bone's outer membrane), the cortical bone, and surrounding soft tissue are ruptured releasing free flowing blood. This blood clots around the break and becomes hypoxic (lacking a oxygen supply), damaged bone tissue loses nutrition and this creates a inflammatory response in which leukocytes (white blood cells) are moved to the fracture site. The purpose of this first phase is to stop infection in the new fracture site. The inflammation phase also brings fibroblasts, stem cells and endothelial cells which bridge the fracture with a loose bond of tissue. The second step in fracture repair is the soft callus phase in which the granulation tissue is replaced by fibrous tissue connecting the break with a soft callus. Next is the hard callus phase in which blood vessels enter the newly formed soft tissue releasing osteoblasts. Osteoblasts are cells that control the passage of calcium within a bone. These osteoblasts convert the soft callus to hard callus, which is made up of of mineralized bone matrix. When the broken bone is connected by a hard callus the final remodeling step begins. In the remodeling phase blood is reverted back to the bone and their bone undergoes bone regeneration until it becomes the same size and shape as before the break.

Bones have a normal upkeep that helps to keep them healthy and that also contributes to bone healing. Bone regeneration is the ongoing replacement of old bone tissue with new bone tissue. Osteoclasts break down old bone matrix and collagen fibers, while osteoblasts create new bone matrix and collagen fibers. This process works to keep bones strong, healthy and is present in fracture healing. Bone regeneration works on newly healed bones to remodel them into the shape they had before the fracture.

Bones fracture in different ways depending on the trauma the bone is exposed to. These different fractures heal differently than others. Bones have four different types of fractures Partial, Complete, Closed and open. Partial fractures are incomplete breaks that crack the bone. Complete fractures are when the bone breaks into two or more pieces, they often require surgery to set the bone back into proper position to heal. Closed fractures do not break through the skin, and heal normally when stabilized and protected, but depending on whether the bone has moved or not, may require surgery. Open fractures are classified as having the broken ends of the bone protruding through the skin. Open fractures run a high risk of infection due to their exposure to bacteria outside the body. Subsequently they require immediate attention and surgery within the first six to twenty-four hours of the break. Open fractures also require a high dosage of antibiotics to fight infection.

The body's process of fracture healing, while highly effective has serious drawbacks. Some fractures cannot heal naturally and require surgical implants to assist in healing. Primary healing can only occur when a bone fracture is stable, the use of splints and cast help to keep the bone in place but sometimes these measures are not enough. Surgery is used to hold bones that cannot be kept in place with a traditional caste or splint. Screws, plates and pins hold the bone together to allow for the healing process to start.

Bones compose the framework of the human body. Bones constantly heal and breakdown, they are strong, resilient and hold many times their weight. When bones fracture they heal in specific phases and sometimes require medical attention. The skeletal system is entirely dependent on bones and keeping them healthy is a major part of health.

**How do T-cells work in the human immune system?*

David Chau

The human immune system is a complex network that functions through the collaboration of multiple organs and cells to keep us safe from disease-causing entities. In this system, T-cells are vital guards of our immunity as they destroy foreign threats and alarm the whole immune system to defend our bodies.

To discuss how different types of T-cells function, we should know about the immune system and the T-cell. “The immune system” is an umbrella term for a set of different types of immune systems, such as external and internal immunity, adaptive and innate immunity, and also humoral and cellular immunity. These taxonomies identify the characteristics of each system. Specifically, the adaptive immunity, in which T-cells are in, is the system that adapts our body accordingly as it encounters different diseases. Unlike innate immune systems, the adaptive system is set up to be exposed to different diseases so that the immunity can responds and be prepared for later attacks. Furthermore, T-cells are one of the two main types of cells in the adaptive immune system, with the other type called B-cells. T-cells bear this name because of their origin, thymus. There are many types of T-cells according to their functions, with the most two fundamental types, which are Helper T-cells and Cytotoxic T-cells. Apart from those two, there are also Memory, Effector, Suppressor T-cells, etc. They all contribute to the physical well-being of the body at the present and in the future.

The first type of T-cell is Helper T-cell, also written as Th. The main function of these cells is to work as the "alarm" of the immune system. When abnormalities are detected inside by Dendritic cells, they activate other T cells for defense. A Dendritic cell delivers a sample of the antigen to a T-cell through its receptors. These receptors vary from cell to cell and only one type of receptor can respond to a certain antigen. Now that the T-cell has a copy of the pathogen's gene, they are activated and start duplicating, each new cell carrying the same information as its mother. Then, these Helper T-cells release cytokine, a protein, to alarm other parts of the immune system to prepare for battle. Moreover, after being activated, some Th cells go on to be Memory T-cells. These cells are as they continue to preserve the genetic information of the antigen for later use. Whenever the same antigen attacks the body, the Memory T-cells will help the body trigger a more efficient defense and terminate the foreign threat before it can harm the body.

Another important type of T-cell is the Cytotoxic T-cell (Tc). They directly attack the foreign threat. Tc target every cell that threatens to the body. Unlike Helper T-cells, Tc can receive the cancerous or infected cells' genes directly from their receptors, which are also original to every cell, instead of having a third party such as Dendritic cells. When a Cytotoxic T-cell receives the information of the infected cell, that T-cell is activated. The cell starts to duplicate itself, just as Th cells. During this process Tc-cells also separate into two types: Memory T-cells, which memorize the genes of the threat for later use, and Effector T-cells that do the killing of the antigen. The most common way for a Cytotoxic T-cell to kill an infected cell is that the T-cell releases Perforin, a type of protein that make holes on the infected cells, and Cytotoxins, which triggers the infected cells to self-destruct and terminate the viruses inside. Therefore, killing the threat.

The human immune system is a complex network of different cells with different functions to protect the body from any foreign disease-causing agents and T-cells play an important role in sustaining the system. Nevertheless, it is the most optimal to live healthy and maintain health to avoid any diseases.

**Why are humans unable to develop immunity against annual allergens, such as mold and pollen?*

Natasha Samsonov

According to the American Academy of Allergy, Asthma, and Immunology (AAAAI), thirty percent of the population worldwide suffer from hay fever and seasonal allergies each year. Although most people have experienced allergies and their effects on the body at some point in their lives, many do not understand what is really causing their symptoms. In order to comprehend how allergies create such an impact on the body and if they can be cured, one must understand what an allergen is, the process and protocol the immune system goes through when introduced to an allergen, the symptoms, and finally how allergens can be put under temporary control.

First, one must decipher what an allergen truly is. An allergen is a typically harmless substance that causes an unwanted allergic reaction in the body. In other words, when the immune system is introduced to a specific allergen, the body misidentifies it as a harmful foreign invader. This is the reason that different people might be allergic to different substances, due to the way their bodies react to these allergens and due to the severity of these reactions. Examples of allergens include pollen, animal fur, mold, dust, and even some foods like nuts and wheat.

Most have experienced the runny nose and sore throat, but one might not understand what is really happening in their immune system during a response to an allergen. First, lymphocytes, or white blood cells, in the immune system detect an allergen and mistake it for a dangerous substance. In order to neutralize the threat, lymphocytes release blood proteins called antibodies. Lymphocytes release a specific antibody called immunoglobulin, or IgE. IgE then attaches itself to immune cells called mast cells. When these mast cells are introduced to an allergen, they overproduce the chemical histamine, which is responsible for causing the common allergy symptoms like a runny nose, watery eyes, and swelling.

Because indications of allergies can be inconvenient and troublesome, most still wonder if their allergies will ever be dissipated or cured. To this day, unfortunately, scientists and doctors have not found a cure to prevent allergies permanently. A common misconception about allergies is that they are a result of a weak immune system, yet it is actually the opposite. Allergies are the result of an overly active immune system, which is the reason one's body does not build an immunity to them. With an overly active immune system, the body continuously tries to neutralize the allergen, leaving one with the tiresome allergy symptoms each year.

Although seasonal allergies are incurable, there is a solution to make indications less noticeable and severe, through the use of antihistamines. Antihistamines are drugs that temporarily relieve allergy symptoms. When antihistamines enter the body, they travel to mast cells and work to block histamine receptors, the sites that release the chemical. Therefore, the secretion of the chemical is reduced. Although antihistamines quickly relieve indications of allergies they are temporary and typically only most effective one to two hours after being taken, until symptoms begin to become more noticeable once again.

As expressed, humans unfortunately have not yet been relieved of their seasonal allergies and trying hay fever. Although the use of antihistamines temporarily alleviate allergy indications, they only last for a short time. It is yet to be discovered if people with overly active immune systems will ever be able to develop immunity against their allergies, but hopefully this discovery will soon be made.

***What is a cancer cell?**
Soo Hyun Choi

Cancer is something that has always been with the human race. The minuscule cancer cell causes enormous damage, which sometimes leads to death. Although that is the basic definition of cancer, the more detailed explanation of this disease has been improved over and over again as time progresses. One of the critical points of this detailed explanation is the start of it, the cause of a cancerous cell.

Viruses can cause a cell to become cancerous. A virus contains RNA or DNA and then injects it into a host cell. Then, the virus will manipulate the host cell's enzymes and make more copies of its own genetic material. Another cause is one's genes, which could contribute to the birth of cancer. Overall, all of these causes do one thing; they cause too many mutations in one's genetic makeup, which makes the oncogene express itself.

All of the items listed previously lead to the expression of an oncogene. First of all, an oncogene is a cancer-causing gene (onco- is Greek for tumor), and before a healthy gene becomes an oncogene, it is called a proto-oncogene. So, an oncogene is a gene with too many mutations in it. Although, this does not mean that when a gene has a mutation in it, that it is an oncogene. When a healthy cell finds a mutation in its genes, it will probably shut down on its own. However, there are certain cases when a cell does not shut down, and the mutations continue to appear. These mutations are what lead to cancerous cells, in which they will malfunction and cause damage to the body.

After an oncogene expresses itself, this leads to a cell malfunctioning. Before this malfunction, then cell divides itself normally at a controlled rate. But, due to the oncogene, the cell will split at a rate that is much faster than the controlled one. This excessive division leads to an issue in the body.

As the cancerous cell malfunctions, the results are the creation of a malignant tumor. A malignant tumor is created from an unusual cell production, which causes a strange lump. This lump holds cancer cells that will invade nearby tissue or bloodstream, which will spread the disease. Although, this does not mean that all tumors are cancerous. For instance, a benign tumor is a tumor that has cells that do not invade nearby tissue. Thus, benign tumors are not cancerous. However, a benign tumor can become a malignant one, but only rarely. Going back to malignant tumors, they can occur anywhere on, or inside the body, since the cancerous cells can attack nearby tissue and go into a nearby bloodstream.

Cancer. Present in the past as well as today, and all from one cell. It is almost astonishing how some mutations in a gene could make a malignant tumor, frightening even. From a list of causes, an oncogene can be born. Then, a cancerous cell spreads itself throughout the body. Finally, a malignant tumor appears. All from one cell. One cell.

The Human Body

**** How do cells communicate with each other, and what molecules are involved?***
Donovan Quinn

The four types of cell signaling are direct contact, the paracrine system, the endocrine system, and the autocrine system. Each one has a unique function in signaling cells in the human body. For each process, there is a different type of signaling molecule used. Usually, signaling cells are called ligands. Ligands are released from a sending cell, often into the extracellular space. For a ligand to trigger a reaction, it must be recognized by its target cell with the right receptor protein. When a ligand binds to a compatible receptor protein, it modifies the activity or shape of the receptor, causing a change inside the target cell. After the message conveyed by the ligand enters the cell, it is relayed through a chain of chemical messengers that eventually causes a change in the cell, called a response.

The first way that cells communicate is through direct contact, where cells can either communicate by bonding complimentary surface proteins, or by sharing signaling cells through a gap junction in animal cells or plasmodesmata in plant cells. These gap junctions and plasmodesmata are called intracellular mediators, which are water-filled channels in cells. These intracellular mediators allow signaling molecules to diffuse between two cells, without ever entering the extracellular area. This type of cell signaling is important because it allows cells to create large areas of connected cells. This makes it so if one cell is signaled by a ligand, it can transfer the message to all of the cells it is connected to by an intracellular mediator. This lets cells coordinate a response to a signal, so they all respond at the same time. The second form of direct contact signaling is through complimentary surface proteins, where cells have proteins that bind to one another. This causes one or more of the proteins to change shape, causing a response.

The second form of cell signaling is the paracrine system, also known as short distance signaling. It consists of a cell releasing ligands for the receptors of nearby cells to recognize. The ligands associated with the paracrine system are called paracrine factors, which can diffuse through spaces between cells. This type of signaling is often used in development, when a cell releases paracrine factors to tell nearby cells which cellular identity to take on.
The third type of cell signaling is the endocrine system, or long distance signaling. This type of signaling occurs when an endocrine gland releases ligands into the bloodstream for the receptor of a cell that is far away to recognize and trigger a response. Endocrine signaling is the most common form of cell signaling, and the ligands used are called hormones. In this process, hormones are released from a cell into the bloodstream, where they move through the body to another cell farther down the bloodstream, which recognizes the hormone and causes a response.

The fourth and final form of cell communication is the autocrine system, which is a different type of system than the others because in this process a cell communicates with itself to trigger a response. Although this doesn't seem like a very practical way to transmit a signal, it is helpful because it changes the structure of the cell when the ligand bonds to the receptor. This form of cell signaling often works with the paracrine system in processes like development.

Overall, each form of cell signaling has a unique function in the human body. All 4 types of cell signaling take place over different distances and with different signaling molecules, but they all work together to connect the cells of the human body. If it weren't for cell signaling, the human body would just be a bunch of unorganized cells.

How do your muscles turn chemical energy into mechanical energy?
Daniel Xu

Muscles are a very important part of the body because they are involved in everything from walking to moving. But how exactly do muscles work? We will investigate how we use our muscles and how we turn the chemical energy we have in our bodies into mechanical energy. Although it may seem complicated, muscles work using just 5 simple steps. First, the brain sends a signal to the muscle which activates the release of calcium ions into the muscle fibers. Then, the calcium ion binds to troponin (a protein) which moves the tropomyosin (a protein) away from the active site of actin (a protein). The myosin proteins then attach to the actin filament forming a cross-bridge. ATP energizes the myosin which pulls the actin filament in, shortening the muscle. The myosin detaches, and finds another actin to pull in. Finally, the signal stops and the calcium ions are removed from the troponin and the muscle returns to a relaxed state.

Step 1: Signal
First, the brain sends a signal to the motor neuron. The structure of the muscle is unlike the structure of other parts of the body. Technically the whole muscle fiber is one big cell with many nucleus's spread throughout the muscle. At the motor end plate of the motor neuron, acetylcholine (a neurotransmitter) is released into the synaptic cleft of the muscle. This depolarizes the motor end plate. The action potential travels through the transverse tubules into the sarcoplasmic reticulum (endoplasmic reticulum of a muscle fiber). During a relaxed state, the terminal cisternae (enlarged area of the sarcoplasmic reticulum) of the sarcoplasmic reticulum is filled with calcium ions; when the signal reaches the sarcoplasmic reticulum, the calcium ions are released into the rest of the muscle. The calcium ions are what triggers the muscle into actually moving, and the stronger the signal, the more calcium is released, and the harder the muscle is contracted.

Step 2: Binding
The myosin proteins to bind with the actin filaments, but the protein tropomyosin is covering the active site of actin. This makes it so that when in a relaxed state, the myosin cannot

interact with the actin and does not cause the muscle to move. The calcium ions then bind to the protein troponin, which moves the tropomyosin away from the active site of the actin filaments. The myosin heads can now bind to the actin forming a cross-bridge. The myosin can now pull and cause the muscle to move. myopathy is when one of these components are defective and the muscle cannot move as strongly.

Step 3: Pull
ATP reacts with Myosin to create ADP and phosphate, which causes myosin to be in a high-energy state. The myosin uses the energy to pull the actin in and shortens the muscle. This is how the muscle can contract. A muscle cannot expand by itself but must rely on the opposite muscle to contract in order to loosen.

Step 4: Release
The myosin then detaches from the actin, and finds another actin filament to pull in. Each contraction shortens the muscle fiber by about 1%, so this contracting action must be done multiple times. The more muscle fibers in a muscle, the more pulling power a muscle has. Generally, half of the myosin is pulling, the other half is finding another actin to pull in.

Step 5: Stop
Acetylcholinesterase breaks down the acetylcholine at the neuromuscular junction (junction between motor neuron and muscle fiber). The sarcoplasmic reticulum stops the release of calcium ions, and the calcium ions are pumped back in through active transport. When the calcium ions leave troponin, the tropomyosin proteins once again cover the active site of myosin, and the muscle returns to a relaxed state.

In summary, muscles work in just 5 simple steps. A signal from the brain is converted to a release of calcium ions into the muscle. The calcium binds to troponin, removing tropomyosin from the active site of actin. Myosin is then able to bind to the actin filaments, and pulls the actin in. The myosin proteins repeat this process until the signal stops. The calcium ions are pumped back into the sarcoplasmic reticulum and the muscle returns to a relaxed state.

What actually happens when someone "pulls a muscle," and how does one recover from it?
Vanessa Stamper

Humans contain all of the muscle fibers they will ever have at birth, so one cannot grow new fibers, only strengthen or weaken them. To strengthen them, one may work out or lift weights. However, if not properly performed under the right conditions, one can end up weakening them. This weakening in the muscle might cause a painful "pull" in the muscle, leading one to come to the conclusion that they "pulled their muscle." Thus, it is important to understand what happens when these fibers that make up the muscles, specifically the skeletal, are abused and how to efficiently recover from it.

Contrary to popular belief, when someone "pulls" a muscle, the correct term to use is a muscle tear. The word tear suggests a structural injury in the muscle fibers that weaken the muscle's efficiency and contractile properties. To be labeled a "tear," at least 5% of the fibers in the muscles must be lost. As a result of overuse, a muscle can tear; most of these tears are due to fatigue-induced muscle disorder. To elaborate, this is when exercise at longer initial muscle lengths induces a larger amount of strain on the muscle fibers. The outcome is a linear deformation of the sarcomeres in the muscle beyond their normal length. Some symptoms may be swelling, bruising, or redness due to the tear, pain at rest and while using that specific muscle, and inability to use the muscle at all.

To understand what happens in one's body when they tear a muscle, you must first understand the structure of the muscle and how it works. The muscle is composed of muscle fibers that are made up of very fine, contractile fibers called myofibrils, which are then divided into sarcomeres. In these sarcomeres are the myofilaments actin and myosin, which cause the contractions of the sarcomeres. This ultimately controls the movement of the muscle. Most tears happen during the contraction period, when the "heads" of the myosin attach to the actin filament, pulling the actin towards the middle of the sarcomere. It is important to remember that muscles never push, they always pull.

To properly recover, one must allow these muscle fibers to thicken and build back up again. To accomplish this, one may follow the "RICE" rule: rest the injured muscle, ice the area to reduce swelling, compress the muscle, and elevate the injured area. It is best to avoid pain medications because they can interfere with the body's normal inflammatory recovery process. If the muscle tear is severe, it is important to watch for bruising, which occurs when the muscle tear is so intense that it bleeds into the body, causing the purple color you see when a bruise develops. Another important part of recovery is not stretching or rolling out the muscle after it tears. This will cause the muscle fibers to "pull" apart even more, increasing the intensity of the tear.

Ultimately, when one "pulls a muscle," they actually tear the muscle, leading to a loss of muscle fibers. The main factor leading to a tear is overuse of these muscle fibers that causes a malformation to the usual length of the sarcomeres. Proper healing of the fibers is vital to ensure that one's muscle recovers to its normal strength. After all, skeletal muscles make up 40% of the body's mass.

How do human bodies respond to stress and danger?

Ethan Liang

Human bodies are able to adapt to dangers and stresses that occur in our lives. The main response that humans exhibit in dangerous or stressful situations is the fight or flight response, where the body's physical abilities increase, improving the body's ability to either fight or flee. A variety of human body systems are involved in this response, including the sympathetic nervous system and endocrine system. While this natural response can help humans survive, frequent stressful situations will damage the body parts involved in the response. The fight or flight response occurs not only when a physical hazard is encountered, but also when one is faced with mental stress. Another, less common response to stressful situations, is the tend-and-befriend response. This is where humans will bond with other humans and support each other.

The sympathetic nervous system and endocrine system work together in the fight or flight response to improve the body functions required in threatening situations. Through the nerve pathways originating from around the middle of the spinal cord the sympathetic nervous system starts several reactions in the body. The adrenal medulla near the kidneys, a type of adrenal gland part of the endocrine system, is signaled by nerve impulses originating from the hypothalamus in the brain to release the hormones epinephrine and norepinephrine, more commonly known as adrenaline and noradrenaline into the blood. Additionally, the pituitary gland, also a part of the endocrine system, releases adrenocorticotropic hormone, or ACTH. ACTH travels in the blood to the adrenal cortex, which causes two to three dozen more hormones that improve body part functions to be released. Together, all these hormones help increase heart rate, blood pressure, blood sugar, muscle tension, and they cause pupils to dilate, and blood to be redirected toward major muscle groups. All of these effects are very useful in either fleeing quickly or fighting strongly; awareness is also increased. Body functions that are not immediately important, such as the digestive and immune systems, are temporarily shut off.

There is an alternative psychological response to stress that occurs in some people, known as the tend-and-befriend response. This response is when one chooses to get close with others in times of stress to lessen their fear. It is a more common response to psychological stress, rather than physical danger, and it has been found to occur more in women due to their tendencies to directly care for children more. When humans interact with those they are close to, endorphins are released from multiple parts of the body. These endorphins can block pain and increase happiness, so by tending and befriending in times of stress or danger, humans are able keep calm and content.

The beneficial body changes that occur in the fight or flight response come at the cost of after effects, which can become very harmful if one has this response frequently. After a period of this fight or flight response, the reserves of these hormones decreases and eventually becomes insufficient. The muscles that are affected by the hormones become damaged, as well as the heart, circulatory system, and brain. The immune system is not fully functional, which was caused during the fight or flight response, so the body is prone to many diseases. All of these effects come after response to any kind of stress, including everyday mental stress. Therefore, it is important for humans to not be constantly stressed or in danger.

The human body has a variety of possible responses, which involve various systems, to stress and danger. There is the common fight or flight response, where the body's physical capabilities are increased due to the release of certain hormones. There is also the tend-and-befriend response, where humans bond with one another to be emotionally uplifted with the release of endorphins. Humans have survived for thousands of years through all kinds of dangerous situations thanks to this ability for our bodies to adapt.

*How do humans process sound with their ears?

Matthew Cao

Humans are constantly surrounded by sound, vibrations in the air. To apply these sounds to their daily lives, humans need to understand them. The human brain is unable to accept these vibrations, but it has the capacity to understand electric charges. The human ear is an organ that, through a long series of steps, translates sound waves into the electric charges.

The first step for sound waves to reach the brain is directing these vibrations through the ear to the middle ear. The vibrations arrive at the ear mostly through air. To "catch" these air molecules that are traveling through the air, the ear has a spiral funnel structure named the pinna that directs these air molecules into our ear canal, which is a tunnel that connects the pinna to the eardrum. The eardrum is a tympanic membrane, which means that it has a triangular, semicircular structure. As air molecules hit off the outer side of the eardrum, three tiny bones located on the inner side start vibrating. These three tiny bones are the malleus, incus, and stapes. The malleus is connected to the eardrum and gets triggered. It acts like a hammer as it vibrates against the incus. The incus, connected on one end to the malleus, links to the stapes. And finally, the stapes brings the vibrations to the oval window.

The stapes pushes a fluid in the cochlea through the oval window to start the translation of vibrations to electric charges. The cochlea is a cavity that is filled with liquids and membranes. As the stapes vibrates back and forth, it pushes the oval window, an opening in the cochlea. The oval window, being pushed by the stapes, pushes a liquid back and forth inside the cochlea. The vibrating fluid makes its way to the end of the cochlea and turns and travels back towards the oval window in an adjacent chamber. As the fluid does this, it pushes the membranes that separates the chambers together.

As two membranes are pushed against each other, it triggers cells between them to send electric charges to the brain. Between the membranes there are small cells called hair cells that play a huge role in hearing. The exposed parts of the cells are the shape of little hairs. The hidden parts of these cells are buried in one of the membranes. As one of the membranes is pushed against the other one, the hairs get triggered on contact. The contact then triggers a reaction in the other part of the cell. Connected to that part of the cell is the auditory nerve. The auditory nerve receives the electric charges sent out by the reactions. The auditory nerve then carries these electric charges to the brain.

In the brain, these charges are received and interpreted, completing the process of hearing. Through a long progress, air molecules were able to be translated into electric charges. It required bones, membranes, organs, and nerves. In the end, the human ear can carry the vibrations, transform them into electric charges, and bring them to the brain.

**How do biological membranes function, and how are they structured?*

Daniel Niculescu

In the human body, or all organisms, for that matter, cellular membranes are extremely important. They surround the cell, allowing only certain objects to pass through. They separate the cell from the outside world and other cells. They keep the cell's form and shape, making sure that all of its organelles and cytoplasm simply spill out. Frankly, without cellular membranes, life as we know it would not be possible. The functions of cellular membranes are crucial to how we know cells to work, and their structure corresponds to them.

Cellular membranes allow for passive transport of substances, or transport without using energy. This occurs through diffusion, or the spreading out of substances into areas they do not occupy. Take, for example, a gas in a closed room. The gas spreads out throughout the room eventually, because there is a lower concentration of it in surrounding areas. However, cell membranes are selectively permeable, only allowing certain substances to diffuse across them. The membrane acts as a filter here. Substances move from areas of high concentration to areas of low concentration, but only substances that the cellular membrane allows for. This is called simple diffusion, or osmosis in the case of water. Gases like oxygen dissolve in water and diffuse across the phospholipid bilayer in simple diffusion, and in osmosis, water diffuses across it.

Cellular membranes also permit for facilitated transport, which aids the substance with crossing the phospholipid bilayer. Two kinds of proteins facilitate the transport of these materials- channel proteins and carrier proteins. Water-soluble substances such as amino acids and sodium typically cannot diffuse across the cellular membrane. This problem is solved with protein channels and carrier proteins. Channel proteins are a sort of passageway, with a specific diameter to only allow certain substances to pass, that goes across the cellular membrane. They also often have "gates" which open and close to only allow some substances inside. Another way of facilitated transport is using carrier proteins, molecules which attach to certain substances and bring them across the membrane. When the protein attaches to the substance, it triggers changes in its own shape, allowing it to pass through the membrane. Some transport proteins don't use energy, in fact, only allowing substances across the membrane that would normally diffuse.

Endocytosis and exocytosis are other methods of transport through cellular membranes. Some objects too large or too charged cannot pass through the cell membrane, so they either enter or exit via these two processes. There are three types of endocytosis. In phagocytosis, the cell membrane engulfs the object and brings it in. There is also pinocytosis, where small vesicles detach from the cell membrane and bring in smaller, dissolved objects. Finally, receptor-mediated endocytosis is where receptors on the outside of the cell accept only certain types of objects surrounded by low-density lipoproteins. In exocytosis, small vesicles are transported to the cell membrane, carrying waste materials. It then melds with the membrane, releasing the contents outside.

The structure of cellular membranes corresponds to its functions. It keeps the interior of the cell away from potentially harmful substances. Also, without a cellular membrane, a cell would not have a shape, as its cytoplasm and organelles would simply spill out of it. The exterior of it also contains cell adhesion molecules, responsible for sticking multiple cells together. The structure also allows for various types of transport, with specialized proteins, a semipermeable composition, and it allows for endocytosis as well as exocytosis.

The functions of cellular membranes are incredibly important to all life as we know it, and their structure corresponds to these functions. From keeping a cell's shape to allowing transport of certain objects to protecting the cell's interior from the outside world, cellular membranes' functions are responsible for the way all organisms are today.

**How do humans thermoregulate in order to stabilize heat gain or loss?*

Ella Michalak

Why does one sweat after a long run, or shiver when it's cold? It is because of the body's need and ability to maintain a stable core temperature, or homeostasis. An example of homeostasis is thermoregulation, which is the body's process of keeping it at a stable temperature. Humans thermoregulate differently than many other animals, as they are endotherms, meaning that they can regulate temperature internally. Humans have many mechanisms in order to stabilize temperature.

Homeostasis is the body's capability to keep a stable balance- a key part in regulating temperature. There are two different groups that regulate body temperature: endotherms and ectotherms. Endotherms can regulate most of their body temperature internally, humans falling under this category. On the contrary, ectotherms depend on external sources for the surplus of their heat. The hypothalamus plays a substantial role in regulating the body's internal temperature. Located towards the front of the brain, it evaluates the temperature of the blood flowing through it, using temperature sensors throughout the body, constantly deciding whether it is too hot or cold. It then stabilizes the blood as needed, sending signals to the body through the nervous system.

The body has numerous mechanisms to keep its internal temperature stable. Goosebumps are one example, which occur when each tiny muscle at the starting of the hair follicle contracts, called a piloerection. This has no benefit to humans because there is not enough hair to create insulation but benefits other animals with more fur. Another significant mechanism of thermoregulation is sweat. When the body overheats, it triggers sweat glands to bring water and salt to the surface of the skin, which evaporates off and cools the body. Humans also have blood vessels, or capillaries, near the surface of the skin that can dilate or contract. When the temperature is too low, the vessels contract, restricting blood flow so that minimal heat is lost through radiation. When the temperature is too high, the capillaries open, increasing blood flow towards the surface of the skin, increasing heat loss through radiation. This is why someone appears flushed after exerting physical force. Fevers are another important mechanism for thermoregulation. A fever is the body's response to a bacterial or viral infection. In order to inhibit the bacteria to continue growing, the body brings the set point up (the stable point where the body keeps temperature) a couple of degrees. This results in an influx of temperature, also known as a fever. This is also the reason one feels cold and shivery during a fever, because the body has not adjusted to the increased set point yet, so it is a lower temperature than needed.

For infants however, it is much harder to obtain a stable core temperature, and they are dependent on adults to stay at a healthy temperature. Although humans are endotherms, babies have a harder time maintaining a stable temperature. A baby's surface area to volume is larger than that of an adult's, so they can lose heat much faster than an adult, almost four times as fast. This is problematic because it is critical to a newborn's life to stay warm. The body temperature for infants depends on the heat transfer between the infant and external environment. Babies do not have developed mechanisms for thermoregulation, and lose heat very quickly, especially through their head. One way to keep a baby warm is through swaddling, which is done directly after birth to preserve heat.

In conclusion, the body depends on thermoregulation to keep a stable temperature within the body. Humans are endotherms, and regulate their temperature internally, needing to maintain a healthy balance. The body has many mechanisms to maintain stability to survive, such as fevers, sweating, and goosebumps.

Why are vitamins important?
Emily Chaika

There are lots of pills and drinks that claim to meet your daily requirement of vitamins, but what is a vitamin? Vitamins are organic compounds. They are also called micronutrients because you only need between 0.01 and 100 mg of them per day, depending on the vitamin.

There are 13 essential vitamins humans need, and they all have a certain function. You need to get essential vitamins from your food. Vitamin A is important for bone growth, your immune system, and keeping tissues healthy. Vitamin C can reduce the risk of cancer and helps make collagen. Vitamin D strengthens bones and teeth. Vitamin E acts an antioxidant, and it helps prevent cells and vitamin A from damage. Vitamin K is essential in blood clotting. There are many B vitamins, and some of them are also coenzymes. Coenzymes help enzymes function. Vitamins B1, B2, and B3 are used to convert food to energy. Vitamin B1, or Thiamine, and vitamin B5, or Pantothenic Acid, are needed for nerve function and healthy blood and brain cells. Vitamin B6 helps make red blood cells and can affect brain function. Vitamin B12 breaks down amino acids and fatty acids. Biotin helps synthesize glucose and break fatty acids. Folate is often taken by pregnant women because it can help prevent brain and spinal birth defects.

Even though you can take a daily vitamin, it is possible to get all the vitamins you need from your food. Highly processed foods are much lower in vitamins than fresh food. Meat is high in vitamin A, B6, and B12. Dairy contains many vitamins like vitamin A, B6, biotin, and B12. Fish contains Vitamin D. Most fruits are high in vitamin C, and leafy vegetables contain vitamin K. Other vegetables in the cabbage family contain Vitamin K and B5. Many vitamins can be found in avocado and other fats, like vitamin E, B3, B5, and B6. Grains contain vitamin B1 and B2. Eggs are high in many vitamins such as B1, B3, B5, and B6. Even though humans need to acquire essential vitamins from foods, bacteria, fungi, and plants make their own vitamins.

Vitamins absorb into your body by dissolving in either water or fat. Water soluble vitamins include B vitamins, biotin, folate, and vitamin C. These are digested and picked up by blood. Excess of these is not dangerous because the excess travels to the kidneys and out through the urine. Too much of a fat-soluble vitamin, though, can be dangerous. Excess is stored in the body in liver and fat cells. These attach to proteins and are carried through the blood. These include vitamin A, D, K, and E. It is possible to accumulate toxic amounts of these vitamins.

On the other hand, not enough intake of vitamins can also have negative effects. Although the daily requirement is very small, many people do not meet it. When you don't have enough of a vitamin, you have a deficiency. This can cause a variety of symptoms depending on the vitamin, but some include brittle nails, bleeding gums, and hair loss. Scurvy is the result of a lack of vitamin C.

Although they are very important, consuming the wrong amounts of vitamins can be dangerous. Even though the quantities required small, vitamins have a huge impact on your health.

How do you digest food, and what are the stages of digestion?

Jeewon Han

It was a chilly evening, and Marggie was feeling hungry. Soon, she and her family decided to eat out. She ordered a rib eye steak and a bowl of salad. After finishing her meal, Marggie wonders, "what happens to my food now?" To answer this question, let's take a journey into her digestive system.

Digestion begins in the mouth. The mouth digests the food mechanically by chewing and breaking the food down into smaller bits. This increases the surface area of the food, which helps to facilitate the next steps of digestion. Chemical digestion, which is the breakdown of ingested food through chemicals, also begins in the mouth. Saliva contains Amylase, which breaks down starch. Once the saliva is mixed with the broken-down food, a clump called a "bolus" is formed. The bolus will move onto the next stages of digestion.

As the bolus is swallowed, it moves through the esophagus and into the stomach. The esophagus sends food down to the stomach through peristalsis, which is a repetition of involuntary squeezing movements by the muscles of the esophagus. When the food enters the stomach, it is met with gastric acid, which breaks down complicated peptide bonds into individual amino acids. Within the stomach, the hydrochloric acid will convert pepsinogen, an inactive form of pepsin, into active pepsin. The pepsin will break down proteins in the bolus. The food will be digested in the stomach for about six to eight hours. By then, the food will become a mixture called "chyme".

The chyme moves into the duodenum, which is connected to accessory organs such as the liver, the gallbladder, and the pancreas. The liver produces bile, which emulsifies fat. However, instead of being directly released into the duodenum, it will be stored within the gallbladder and then released at need. The pancreas produces pancreatic juice, which lowers the acidity of the chyme. Also, it contains various enzymes such as proteases, lipase, amylase, peptidase, and nuclease. Each enzyme digests a different type of molecule.

After passing through the duodenum, the chyme then moves into the small intestine which absorbs the nutrients within the mixture. The small intestine is lined with millions of finger-like projections called the villi and microvilli which significantly increase the lumen's surface area available for absorption.

The absorbed nutrients flow into the bloodstream and are delivered to cells of the body. The capillaries, which are within each villi receive the nutrients that diffuse through. However, lipids will be diffused to the lymph vessel, which also runs beneath the villi. As the blood and the lymph fluid flows through the body, it will deliver the required nutrients to each cell, allowing them to function. The leftover material from the small intestine will move onto the large intestine, which reabsorbs the water that was used for the digestion. The remaining waste will be stored in the rectum in the form of feces, which will come out through the anus.

Once the leftover materials are excreted from the body, the digestion of Marggie's dinner is complete. She is ready for another meal!

The Environment

How is global warming affecting different species around the world?
Katherine Li

What really is global warming? Global warming is caused by an increase in greenhouse gases, including gases like CO_2 and methane. These gases cause too much heat to be absorbed. Additionally, fossil fuels like coal, oil, and natural gas are also contributing to the increase of heat being absorbed. The increase of heat being absorbed is what causes the warmer temperatures. Furthermore, CFC's, a gas in items like refrigerators, reflect chlorine atoms, which when reacting with ozone (O_3) causes the it to be reduced into O_2. This causes destruction to our ozone layer, (that reflects the heat) and so it contributes to the rising temperatures. Overall, many different types of gases contribute to global warming, which is causing a rise in temperatures. Because of the higher temperatures and other factors that will be mentioned, global warming is devastating to animals.

Global warming causes the destruction of animals' habitats. A recent study shows that if global warming continues, it is likely that 30% of species will become extinct soon. It has already caused many forest fires, droughts, and other natural disasters, due to the earlier arrivals of spring and longer dry seasons. From these forest fires and droughts, many animals are left with nowhere to live. For example, one study shows that already 70 frog species have become extinct after losing their habitats to global warming and 20 more have disappeared. Additionally, because many animals cannot live under warmer temperatures, many animals have been forced to move from their original habitats. A study shows that some species have shifted territories almost 17 kilometers north, and those in the mountains have moved 11 meters higher.

How does global warming affect marine animals? Global warming in the Arctic is causing the depletion of ice, which increases the risk of polar bears drowning and decreases the amount of food that is available. A study shows that 11 of 17 penguin species are also at risk of extinction. Various species of salmon in rivers are finding it difficult to return from their spawning grounds as the flow of many rivers are decreasing. These salmon species up these rivers, back to their original habitats, but without this flow, many cannot return to spawn. Global warming is also causing flooding in rivers, which is causing fish eggs to float away and brings in toxic material into the rivers. The higher temperatures also result in a higher number of female sea turtles and a decrease in male turtles since male turtles are only birthed in cooler temperatures.

The effect of global warming on land animals is very similar to the effect it has on marine animals. For example, it is causing pandas to go extinct, as pandas' natural habitat is in the mountains where it is cooler. Since there is not an abundance of bamboo high up in the mountains, pandas can only stay where they are. Furthermore, the American pika is also greatly impacted. The pika needs to be in an environment that cannot exceed 77°F, or else they would die. The rise of temperatures is causing many pika species to disappear in several regions in the US.

Global warming is causing many animals to be forced in very selective situations. Not only do animals have to compete for food and shelter, evolution is also in need of occurring much faster. One example of the need for evolution is the snow owls. Some animals that have also adapted to global warming are; owls, snapping turtles, 2-spotted ladybugs, and many more. Even though the urgent need for evolution seems quite simple from shown above, but to many species, evolution in such a fast pace is impossible. Since many animals are yet to adjust, this causes a timing error/separation between species that have and haven't reacted. Thus, a decline in population occurs for many species.

In conclusion, global warming has detrimental effects on many animals, as it destructs not only their habitats but their way of life. The rise in temperatures, sea level, and many more caused by global warming, is causing havoc for the environment/habitats of many animals. Additionally, global warming is causing an increase in the need of evolution, which is resulting in the lowering population for some animals.

In what ways are humans responsible for climate change?
Gjulia Camaj

Climate change can be attributed not only to natural factors in the environment, but also to human's mass production of heat-trapping gases. On earth, the balance between the Sun's light and the atmosphere work together to maintain sustainable temperatures for humanity. However, over hundreds of years, the climate increase has caused a lot of concern for the future. Not only are temperatures on the rise, but gases are also associated with the trapping of heat in the lower atmosphere. The production and increase of these gases are linked to human activity such as industrialization.

The atmosphere and the sun work together to create a "greenhouse effect" to control climate on Earth. Half the sun's light reflects off the ozone layer (albedo) while the other half passes through the atmosphere and clouds to be absorbed by Earth's surface. Afterwards, the remaining heat is radiated off the earth infrared heat. Greenhouse gases then absorb about 90% of this heat and release it throughout the lower atmosphere which in turn warms the earth because this heat becomes trapped. This process occurs because molecules in water vapor move faster after absorbing heat which respond to greenhouse gases such as carbon dioxide, methane, and nitrous oxide by not being able to radiate the heat out of the atmosphere.

Various gases created by human activity contribute to heat and infrared radiation being trapped in the atmosphere, raising the temperature of the earth. Carbon dioxide is one of these heat trapping gases that can be found in high concentrations in the

Earth's atmosphere. Levels of Co2 have been on a rise since 1959 due to the burning of fossil fuels and deforestation. Nitrous Oxide is another greenhouse gas that is produced because of soil cultivation, biomass burning, etc. Lastly, methane is another major heat-trapping gas accumulating in the atmosphere because of waste and agriculture decomposition in landfills. Although there is an abundance of heat trapping gases, there are only a few that have caused a spike in climate change. These greenhouse gases need each other in order to trap heat. For example, Mars' thin atmosphere lacks water vapor and methane. Without methane or the quick moving molecules of water vapor to be trapped by C02, the surface of Mars is not warm enough to sustain life.

A leading question still stands on who should be held more responsible for climate change, humans or the environment. Greenhouse gases absorb heat in the atmosphere and distribute it throughout the lower parts of it. The average of the sun's energy has only seen a constant or slight increase since 1970. Temperature rise would be observed in all layers of the atmosphere if the sun was becoming more effective. Greenhouse gases continue to trap heat in the lower atmosphere, in contrast to cooling in the upper atmosphere. In conclusion, humans are the main contributor to the globe warming because of the spike in greenhouse gas production over many decades

How can algae provide a renewable energy source?
Justin Liang

Imagine a world where the air is clean and global warming ceased to exist. This is a world where algae is used as fuel.

There are several different methods for using algae as a fuel source. Algae uses photosynthesis so efficiently that they can double their weight many times a day. Algae accounts for almost 50% of the world's photosynthesis. Algae produces oil which can be used as fuel. The oil they produce can be obtained by using sound waves known as solvents to shatter the cell structure of algae. However, this oil cannot be used right away. It first must be processed at an integrated biorefinery. Single-celled microalgae can also be utilized by extracting the fats they produce, which can be made into biodiesel.

The main advantage of utilizing algae is the fact that they are extremely efficient compared to the amount of resources required in what they produce. Single-celled microalgae produce high quantities of fats while the former produces large amounts of oil. Another important benefit of algae is that it pulls in carbon dioxide and combines it with water to make carbohydrates. Algae pulls in more carbon dioxide than any other plant for its high productivity. This helps limit global warming drastically. In fact, if there was an algae farm next to every power plant in the world, global warming could be solved!

Before the vast benefits of algae can be utilized, there are some obstacles that must be solved. One problem is finding a solution to optimizing the nutrients, light supply, and CO2. Another major problem is trying to bring the price down to 1 to 2 dollars a gallon. The current price is 8 dollars a gallon. Algae not only produces an energy rich oil but are also rich in nutrients. If we can learn to extract this, the price of its oil could be brought down a good margin.

Corn and soybeans have been used as biofuel in the past, but algae provides several advantages over them. First, Algae produces and generates oil 15 times more per acre than any other plants (eg. corn) that have been used as fuel sources. Furthermore, for every 1000 acres of algae, that equates to 40,000 acres of crops that don't need to be planted. This also helps against deforestation which other plants do not. Algae also helps environmentally against global warming. Growing algae next to power plants this allows for the algae to capture the carbon dioxide and prevent it from contributing to global warming. This carbon dioxide also improves the productivity of the algae.

Algae farms are already being experimented with. A company called Sapphire Energy has a total funding of over 300 million dollars with 100 million dollars of the investment being provided by the U.S. government. The company has used the money on a 300-acre algae farm. This company uses the method of extracting oils rather than fats. Another company called Algenol Biofuels has 850 million dollars invested to build an algae farm that can produce an ethanol fuel for the price of 3 dollars a gallon. The company estimates their process will be able to produce around 6000 gallons per acre per year. The company Solazyme provided hundreds of thousands of gallons of fuel to the U.S Navy and Air Force through algae and other biofuels in 2010. Using algae as a renewable source of energy provides many advantages over other types.

How do mycorrhizal fungi contribute to ecosystem function and survival?*

Rachel Faust

Mycorrhizal fungi are critical contributors to ecosystem function. By creating a complex underground communication and transport system, they help ecosystems thrive. Mycorrhizae can buffer threats of erosion, plant infection, and drought. They are also able to increase the survival rate of tree seedlings by four times, contributing to the overall growth of a forest and enhancing the survival of other organisms in an ecosystem. Mycorrhizae are able to function efficiently because of their structure.

In many aspects, mycorrhizae are similar to ordinary fungi. They have a net of small, threadlike tubes, called hyphae, that form an underground structure called a mycelium. The mycelium functions much like the roots of a plant, absorbing nutrients from the surrounding soil. However, mycorrhizal fungi exist entirely underground, and their mycelium is enormous, sometimes covering square miles of soil. Their mycelium penetrates the root structures of plants, exchanging nutrients and forming mutualistic symbiotic relationships. Mycorrhizal fungi are present in every major terrestrial biome and are the most common and widespread examples of symbiosis.

Plants benefit from this relationship in several ways. They can receive nitrogen, phosphorus, iron, and other essential nutrients through the mycorrhizae. Because their mycelium is so extensive, mycorrhizae expand the plant's range and surface area for absorption hundreds of times, allowing for maximum nutrient intake. Mycorrhizae also secrete enzymes that denature some of these nutrients, making them more easily available to plants. The enzymes also destroy certain pathogens, increasing the plants' chances of survival.

The absorption of mycorrhizae contributes to soil health. By making the soil more absorbent, they control erosion, and can improve water quality. Absorption is also helpful to the environment in damp climates, where stagnant water just below the surface of the soil can breed harmful bacteria, a process called eutrophication, potentially creating a higher risk of infection for the plants. The mycorrhizae absorb this water, which can then be transported through the hyphae to a dehydrated plant. Mycorrhizae also keep the concentrations of various nutrients balanced, by absorbing them in areas of high concentration and transporting them down their concentration gradient to places where they are scarce. Nutrient loss is a major threat to ecosystems. By transporting essential nutrients directly to plants in need, mycorrhizae keep the nutrient cycle more closed.

Mycorrhizae also form complex two-way connections between individuals belonging to different species. This network allows different species to be symbionts of each other. They use the mycorrhizae as a means of transport, through which they trade essential minerals and nutrients, as well as defense signals. In addition, studies have proved that parent trees favor their own seedlings, sending defense signals and essential nutrients to their offspring first. This special protection is a critical factor in the survival of seedlings in the competitive world of plants. By making more plant material available for herbivores, and, consequently, more food for carnivores, mycorrhizae support an ecosystem on multiple levels.

Mycorrhizae assist ecosystems in many ways. By providing plants with a method of communication and transport, they support the ecosystem in a way that few other organisms can. Plants with mycorrhizae are more resistant to disease, drought, and ultimately, death. Therefore, mycorrhizal fungi play a vital part in the health and sustainability of the world's ecosystems.

***What is ocean acidification, and how will it affect marine organisms?**
Nicholas Juzych

Ocean acidification, the evil twin of global warming, is a major threat to organisms living in the ocean by lowering pH levels across the entire ocean and using up calcium carbonate. Ocean acidification is considered the evil twin of global warming because it is caused by many of the same molecules that cause global warming, mainly CO2. The CO2 that is trapped in the atmosphere gets absorbed into the ocean and mixes with water, creating carbonic acid. The pH level across the ocean is being lowered by the acid that is being created in the ocean. Ocean acidification is an issue that needs to be acknowledged same way as global warming because it is just as harmful.

Although ocean acidification and global warming are similar, they have distinct differences. Both of them are enormous issues currently and are being looked into daily, but ocean acidification is not being brought into the public view. We are not doing things that help stop ocean acidification. Global warming and ocean acidification are both caused mainly by CO2 in the atmosphere. The similarities end there, rather than trapping the heat like a greenhouse, the ocean absorbs the CO2. It is lowering pH levels in the ocean while global warming warms the earth. Ocean acidification lowers pH levels, which in turn lowers metabolic and immune response. Although both being caused mainly by CO2 they are harming the earth in different ways.

Carbon dioxide that is being absorbed into the ocean is raising pH levels throughout the entire ocean, this is greatly impacting life in the ocean. A lower pH has the ability to lower metabolic and immune response, this would cause many organisms to be slower in every way. They would have a slower response to diseases and other things trying to kill them or catch them. pH begins to lower when CO2 is mixed with water, its product is carbonic acid (H2CO3) which is a weak acid. When it loses 2 H+ ions it forms carbonate, the released hydrogen ions decrease the pH of water. A lower pH level in the ocean will affect the solubility of and toxicity of chemicals and heavy metals in the water. However, this can operate in both directions depending on the current pH level. At a higher pH, this bicarbonate system will shift and CO_3 will pick up a free H ion. Although this reaction does not do very much as H_2CO_3 has a low solubility constant. As CO_2 levels increase around the world the more dissolved CO_2 also increases, the amount of H_2CO_3 will increase in the ocean, which decreases the pH levels. This is having a very large effect on organisms living in the ocean.

When water becomes saturated with CO_2, it not only reduces the ocean's pH but also lowers the calcium carbonate sources as well. Calcium carbonate, $CaCO_3$, is necessary for building corals, shells, and exoskeletons for many aquatic creatures. As CO_3 levels decrease, it becomes more difficult for marine creatures to build their shells. Thus, as CO_2 levels increase the availability of CO_3 decreases which in turn is reducing the amount available for shell and coral building. This will not allow microscopic organisms to make shells causing a decrease in their population, possibly causing mass extinction for larger species living in the ocean, which will ultimately lead to less food for Humans.

Ocean acidification is an issue that is affecting all life in the ocean, whether it be through lowering pH levels or not enough calcium carbonate. Although it is like global warming it is still completely different in how it does harm. It is killing organisms through lowering pH levels, or not enough calcium carbonate. Calcium carbonate will not allow smaller organisms to create shells or exoskeletons to protect themselves. This would lead to the rapid death of many microorganisms which eventually will not allow larger species to feed. Ocean acidification is a very serious issue that can change the way we live if we don't do anything about it.

Assorted Topics

**How does ethology study the way animals learn, live and survive?*
Ben Angileri

Ethology is the science of animal behavior. How animals learn, how they act around enemies and how they survive. There are many levels of learning for animals, from birth to death animals continue to learn and problem solve. By marking their territory, they create their own private area, very valuable to them. During fights certain animals act in a flouting way. All of these behaviors lead to the way animals act around each other.

When studying animals, there is a wide continuum of the way certain behaviors are learned, spanning from innate to learned behavior. Beginning with instinct behavior, instincts are something you are born with, which are genetically programed into the animal. For example, when a turtle hatches from its egg it immediately races to the water. This is the turtle's instincts telling themselves to get water. Next is associative learning, which is when you associate one stimulus to one affect. For example, if you have a dog, and you ring a bell, you also must give the dog a treat. Repeating this over and over makes the dog associate the sound of the bell to receiving a treat. With due time, once the bell is rung, the dog will almost always come to you regardless if you have a treat or not. Another learning method is trial and error. This type of learning is when you teach animals complex actions by having them associate behaviors with consequences. Through this, they learn which actions produce pleasant or unpleasant consequences. Next is habituation which is when the subject receives the same stimuli over and over again and they eventually learn to ignore it. For example, if you take a sea anemone and feed it a mussel, they'll eat. Doing this several more times will accustom the anemone to the mussel and when you try to feed the anemone plastic, it will also try and eat it because you familiarized it with the mussel.

The final and most complex behavior is insight. Insight is the animal's ability to problem solve and make decisions. Not all animals have this capability but the most common animal to see this in is the chimpanzee.

A territory is usually a fixed location that an animal claims, in which they exclude other unwanted animals. They are used for feeding, mating and raising youth. If other animals try intruding in on another animal's territory, that animal will try to fight off and scare the others away. It is immensely difficult to maintain a territory that is expansive. Certain animals must not make their territory too large, unless they would like to get overtaken by a foe. There are many ways for animals to establish their territory. Some animals must announce their territorial rights. For example, birds must constantly sing their melodies and squirrels with their quiet chattering. Other animals indicate their territory by scent. The male coyote marks its territory with urine. Once smelled, intruders will often flee avoiding a potentially deadly altercation.

Conflicts between two animal that arise over shortages of food, mates or territories are settled by agonistic behavior. Threats, rituals and combat determine the winner of the subject. The two animals compete in tests of strength, posturing and symbolic displays. In addition, they try to appear larger and more aggressive to determine the champion. One animal usually surrenders. If violent combat might lead to injuring both combatants, they wrestle instead of biting each other. Once two animals settle a dispute, there is rarely any more conflict between them.

How do you build an organ?

Allison Lacy

Scientists are currently developing many different ways to "build" an organ. Organs are very complex structures that are difficult to replicate, and scientists have taken several different approaches. They are currently developing methods to modify animal organs to fit humans, grow organs completely in labs, and to 3D print organs.

The success of tissue engineering hinges on the precise translation of the operation and management of the parts that make up the tissue. To maintain this precise translation, scientists use something called a scaffold. The National Institute of Biomedical Imaging and Bioengineering describes the scaffold of a tissue as “A structure of artificial or natural materials on which tissue is grown to mimic a biological process outside the body or to replace a disease or damaged tissue inside the body.” With this relatively new ability to grow tissues artificially, scientists are now working on growing organs in a lab.

The structures of organs are difficult to recreate considering their many small yet essential parts, which makes using a “pre-assembled” scaffolding appealing to scientists. Pigs’ bodies are surprisingly similar to human bodies, making their hearts prime candidates for the natural scaffolding of artificial human hearts. Scientists use a detergent to dissolve the cells from a pig’s heart, leaving the basic structure of the heart there to be used as a scaffold. Adult stem cells are then put into the scaffold, and they “build” a human heart around it. This works because the stem cells used are essentially cells that can turn into any type of cell required. They are the cells that make up an embryo before the baby forms, later turning into the specific cells needed to make up the human body. Hearts created this way can beat, but they are not yet fully operational.

As there is not yet one infallible way to create an artificial organ, scientists are developing multiple different strategies, such as building artificial scaffolds, in hopes that one technique will eventually prove successful. Anthony Atala, the director of Wake Forest Institute for Regenerative Medicine has created the “first lab-generated human organs implanted in people” alongside his partners. They take a little part of the patient’s bladder and separate the muscle cells from the cells lining the urinary tract called urothelial cells. They then cause these cells to multiply until they have enough to build a new bladder. They put the muscle cells on the outside of the scaffold. Two days later, they put the urothelial cells on the inside of the scaffold. The new bladder is put into an “incubator” that mimics the body, allowing the cells within the bladder to grow together. Eventually, the bladder is put into a person, and the scaffold dissolves. This is just one of the many strategies for building artificial scaffolding that is well on its way to success.

Another method scientists are working on to create artificial scaffolding for organs is by 3D printing them. Fundamentally, a printer puts "laser-guided droplets of cells and scaffold material" onto a moving surface that descends as more and more cells are printed on top until a tissue has been printed. Atala's lab has used such a printer to build a "two-chambered mouse size heart in about forty minutes," as well as kidneys implanted into animals that can successfully produce urine. Other scientists have printed "ear, bone, and muscle structures" that have proven successful when implanted into animals. These experiments implicate that, in the future, 3D printing may able to yield fully functioning organs.

Scientists have not yet found any technique to tissue engineering that will definitely yield success in every attempt to "build" an organ. The structures of many organs are complicated and require different approaches to the scaffolding based on their complexity. So far, the most promising approaches are using the scaffolding of organs from animals, growing organs and their scaffolding in a lab, or 3D printing organs with their scaffolding.

****How and why does the *Mimosa pudica* close its leaflets?***

Emily Zhang

The *Mimosa pudica* appears to sense touch. Their movement is actually the result of how the plant responds to stimuli, touch in this case. What is the *Mimosa pudica*? While this name may be unfamiliar, the plant is more commonly referred to as the touch-me-not or the sensitive plant. Regardless of its name, there is a unique trait that all *Mimosa pudica* possess which sets them apart from other plants. That trait is its ability to close its leaflets while being touched. Although it may seem like a plant that can move by magic, the *Mimosa pudica* is actually a plant that is induced to move due to biological reasons.

Before understanding how the *Mimosa pudica*'s movement occurs, one must first identify the different parts of the plant involved in this process. For example, being familiar with the pulvini is essential. Pulvini (singular, pulvinus) are enlarged bases located on different parts of the plant. They act as a hinges or joints when the plant moves rapidly. There are three different types of pulvini: the main, secondary, and tertiary pulvinus. The main pulvinus is located between two petiole or stalks, as they branch out. The intersection base between two sets of leaflets is called the secondary pulvinus. Lastly, tertiary pulvini are located between a single leaf and a petiole and are the most active in the process of the *Mimosa pudica*'s movement. Each tertiary pulvinus comes in pairs due to how each leaflet is aligned along the petiole in pairs. Furthermore, the tertiary pulvinus acts as a motor organ in the process of the *Mimosa pudica*'s rapid leaf movement. This is due to motor cells or pulvinus cells found in the pulvinus.

The *Mimosa pudica*'s unique turgidity (pressure within a cell) is what provokes the closing of its leaflets. All plant cells contain either a vacuole or vacuoles in its cytoplasm. These membrane enclosed sacs are filled with fluid and used to store nutrients. Their purpose is to maintain a high turgor pressure so that the plant is able to keep its structure. When the *Mimosa pudica* contains a relatively high turgor pressure, the vacuoles of the pulvinus cells swell, causing the leaves to be able to remain upright. In this case, the cells are turgid. However, when the cells experience touch, Potassium or K+ ions are released from the cell. Therefore, there is a higher concentration outside the cell, compared to the inside. Water then diffuses out of the cell through osmosis, a process in which solvents dissolve into an area of lower concentration, therefore preserving osmotic balance. As water diffuses out, the cell becomes flaccid or weak causing the leaflets to close. Only when the K+ ions are regained, will the cells become turgid again, allowing the leaflets to open back up.

The *Mimosa pudica* is able to "feel" seismonastic movement due to electrical signals. Seismonastic movement refers to movements brought about by stimuli or touch. What the *Mimosa pudica* experiences is thigmonasty, or plant movement in response to touch. When the leaflets of the plant are touched, electrical signals from stimuli are transferred into the plant. This touch causes an action potential which is a change in electrical signals that then travels to the pulvinus. As the plant is touched, electrical stimulation passed down through each pulvinus. This is also why touching a single leaf may also cause others to close up. Finally, this electrical signal is what stimulates the change in K^+ ions located in the motor cells. As a result, electricity can also power the *Mimosa pudica*'s rapid leaflet movement.

Scientists and botanists have numerous theories for the purpose of plant movement in the *Mimosa pudica*. Some botanists claim that the closed leaflets are used to protect from potential predators. In other words, when a predator touches the leaflets by accident, the *Mimosa pudica* closes them, hoping to appear less appealing and tasty. Other researchers have related the *Mimosa pudica*'s movement to environmental factors. These include abiotic factors such as rain and heat which could pose a danger to the *Mimosa pudica*. Scientists have also observed the *Mimosa pudica*'s behavior of closing its leaflets at night in addition to when being touched. This led scientists to hypothesize whether or not *Mimosa pudica*'s movement could be related to photosynthesis and optimize energy production.

As one can see, the *Mimosa pudica*'s ability to respond to touch is complex and profound to understand. The motor cells located on the tertiary pulvinus along with the turgor pressure in them cause the plant to move and respond. Electrical signals from stimuli are transported to the pulvinus leading to the change in turgor pressure. Although there are only theories for why the *Mimosa pudica* responds to stimuli, one can see the importance of the *Mimosa pudica*'s ability to move.

**What must be considered when classifying an organism in the different taxa?*

Nam Duong

What makes a cheetah a relative to a cat and why is it more closely related to a cat than a dog? Scientists rely on taxonomy, the science of classifying animals, to answer this question. To distinguish between two species, scientists have kingdoms and orders that sort all types of natural things. This art of taxonomy has a long history and many methods to differentiate between species.

Long ago, the first taxonomy was done by a scientist named Carolus Linnaeus. In the 1700s, Carolus Linnaeus wanted to change the ways of botany. He published a book called "Systema Naturae". In this book, he split all living things into three different kingdoms: stones, plants, and animals. He then divided each kingdom into smaller subclasses. Carolus Linnaeus's work left 2 important long-lasting impacts: the hierarchical classification system and the binomial nomenclature naming method. For this reason, many people see him as the father of modern taxonomy.

When being classified, animals are assigned taxa and a scientific name. The name for a specific species is called Binomial Nomenclature. As the name states, there are two parts to the name. The first part of the name, which should be capitalized, is the genus of the species, and the second part is the specific name of the species. While the first part of the name, the genus, can be part of many different names, because many species can belong to the same genus, the second part of the name is specific to the species.

Looking alike is a sizable factor when classifying an organism. The first type of taxonomy is called the Phenetic System. The Phenetic System focuses on the similarities between species to distinguish and sort them. These similarities can either be external or internal. They can be similarities between skin, teeth, paws, or how many similarities there are in the microscopic levels. In this type of system, felines are a good example because they share similar characteristics and body types. These characteristics can be chemical, anatomical, physiological or ecological. Scientists also look at organisms based on their origins and ancestors. This taxonomy method is called Phylogeny or Cladistic System. With the Cladistic System, the scientists worry more about the ancestor and blood of the species rather than the physical traits. Using the Cladistic System, scientists look for species with similar ancestor and evolution. Some animals evolve drastically over the millions of years and sorting them would be hard if we did not know it's ancestors. For example, without the Cladistic System, people would never have guessed that birds are closely related to dinosaurs.

In conclusion, taxonomy is not easy, and even something as simple as naming a species can be hard. With the vast amount of species living on earth, it is a challenge to find a unique name for a species and to sort it into the system to be easily comparable and organized. Starting with Carolus Linnaeus, scientists around the world found different ways to do sort the organisms and the method evolved as time continued forward. Now, taxonomy is easy to learn and understand, while still having a complex structure to support the many natural things on earth.